Waste & Recycling

Series Editor: Cara Acred

Volume 315

Independence Educational Publishers

First published by Independence Educational Publishers

The Studio, High Green

Great Shelford

Cambridge CB22 5EG

England

ISBN-13: 978 1 86168 765 4

Printed in Great Britain

Zenith Print Group

Contents

Introduction

WASTE & RECYCLING is Volume 315 in the **ISSUES** series. The aim of the series is to offer current, diverse information about important issues in our world, from a UK perspective.

ABOUT TITLE

Recent figures suggest that the UK produces around 200 million tonnes of waste each year. With the global population increasing, what is the solution to our waste problem? This book examines UK, European and global waste statistics, looking at why some cities are better at dealing with waste than others. It also looks at the issue of microplastics, considering how they affect our oceans, fly tipping, electronic-waste and food waste. In the second chapter, which deals with tackling waste, topics such as landfills, incineration, recycling and innovative waste solutions are explored.

OUR SOURCES

Titles in the **ISSUES** series are designed to function as educational resource books, providing a balanced overview of a specific subject.

The information in our books is comprised of facts, articles and opinions from many different sources, including:

⇨ Newspaper reports and opinion pieces

⇨ Website factsheets

⇨ Magazine and journal articles

⇨ Statistics and surveys

⇨ Government reports

⇨ Literature from special interest groups.

A NOTE ON CRITICAL EVALUATION

Because the information reprinted here is from a number of different sources, readers should bear in mind the origin of the text and whether the source is likely to have a particular bias when presenting information (or when conducting their research). It is hoped that, as you read about the many aspects of the issues explored in this book, you will critically evaluate the information presented.

It is important that you decide whether you are being presented with facts or opinions. Does the writer give a biased or unbiased report? If an opinion is being expressed, do you agree with the writer? Is there potential bias to the 'facts' or statistics behind an article?

ASSIGNMENTS

In the back of this book, you will find a selection of assignments designed to help you engage with the articles you have been reading and to explore your own opinions. Some tasks will take longer than others and there is a mixture of design, writing and research-based activities that you can complete alone or in a group.

Useful weblinks

www.bristol.ac.uk

www.theconversation.com

www.discoversociety.org

ewaste.ee.washington.edu

www.environmentenergies.co.uk

www.esrc.ac.uk

www.food.gov.uk

www.forgeskiphire.co.uk

www.fullfact.org

www.grida.no

GOV.UK

www.greenpeace.org

www.theguardian.com

www.huffingtonpost.co.uk

pubs.iied.org

www.independent.co.uk

www.parliament.uk

www.recyclenow.com

www.wastedisposalhub.com

www.weforum.org

FURTHER RESEARCH

At the end of each article we have listed its source and a website that you can visit if you would like to conduct your own research. Please remember to critically evaluate any sources that you consult and consider whether the information you are viewing is accurate and unbiased.

Which countries produce the most waste?

The amount of urban waste being produced is growing faster than the rate of urbanisation, according to the World Bank's report *What a Waste: A Global Review of Solid Waste Management.*

By 2025 there will be 1.4 billion more people living in cities worldwide, with each person producing an average of 1.42kg of municipal solid waste (MSW) per day – more than double the current average of 0.64kg per day.

Annual worldwide urban waste is estimated to more than triple, from 0.68 to 2.2 billion tonnes per year.

The top producers of MSW were small island nations, including Trinidad & Tobago (14.40 kg/capita/day), Antigua and Barbuda (5.5kg) and St Kitts and Nevis (5.45kg), Sri Lanka (5.10kg), Barbados (4.75kg), St Lucia (4.35kg) and the Solomon Islands (4.30kg). Guyana (5.33kg) and Kuwait (5.72kg) also scored highly.

The worldwide average is 1.2kg.

New Zealand (3.68kg), Ireland (3.58kg), Norway (2.80kg), Switzerland (2.61kg) and the United States (2.58kg) were the top five producers in the developed world.

The countries producing the least urban waste were Ghana (0.09kg) and Uruguay (0.11kg)

The World Bank defines municipal solid waste as including "non-hazardous waste generated in households, commercial and business establishments, institutions, and non-hazardous industrial process wastes, agricultural wastes and sewage sludge. In practice, specific definitions vary across jurisdictions."

20 August 2016

⇨ The above information is reprinted with kind permission from the World Economic Forum. Please visit www.weforum.org for further information.

© World Economic Forum 2017

UK statistics on waste

⇨ The UK recycling rate for 'waste from households' was 44.3 per cent in 2015, falling from 44.9 per cent in 2014. This is the first time the rate has fallen since it began in 2010, though the 2015 figure still represents the second highest annual value on record. There is an EU target for the UK to recycle at least 50 per cent of household waste by 2020.

⇨ UK Biodegradable Municipal Waste (BMW) sent to landfill has continued to reduce and in 2015 was 7.7 million tonnes. This represents 22 per cent of the 1995 baseline value. There is an EU target to restrict BMW landfilled to 35 per cent of the 1995 baseline by 2020. The UK comfortably met interim targets for 2010 and 2013.

⇨ The recovery rate from non-hazardous construction and demolition waste in the UK in 2014 was 89.9 per cent. There is an EU target for the UK to recover at least 70 per cent of this type of waste by 2020.

⇨ UK generation of commercial and industrial (C&I) waste was 27.7 million tonnes. This has fallen from 32.8 million tonnes in 2012.

⇨ The UK generated 202.8 million tonnes of total waste in 2014. Over half of this (59.4 per cent) was generated by construction, demolition and excavation, with households responsible for a further 13.7 per cent.

⇨ Of the 209.0 million tonnes of all waste that entered final treatment in the UK in 2014, 44.5% was recovered (including recycling and energy recovery). The proportion that went to landfill was 23.1 per cent.

⇨ In 2014, 64.1 per cent of UK packaging waste was either recycled or recovered compared to 72.7 per cent in 2013. The 2014 EU target was for the UK to recycle or recover at least 60 per cent of packaging waste.

Waste from households

'Waste from Households' is the agreed harmonised UK measure used to report household recycling to comply with the Waste Framework Directive (2008/98/EC). Under this Directive the UK and other EC Member States must meet a target to recycle 50 per cent of 'household waste' by 2020. The UK currently defines 'household waste' using the 'waste from households' measure.

⇨ The UK recycling rate for 'waste from households' was 44.3 per cent in 2015, falling from 44.9 per cent in 2014. This is the first time the rate has fallen since it began in 2010, though the 2015 figure still represents the second highest annual value on record. There is an EU target for the UK to recycle at least 50 per cent of household waste by 2020.

Biodegradable municipal waste (BMW) sent to landfill

The Landfill Directive (1999/31/EC) aims to prevent or reduce as far as possible negative effects of landfilling waste, in particular on surface water, groundwater, soil, air, and on human health by introducing stringent technical requirements for waste and landfills. Biodegradable waste decomposes in landfill to produce methane, a potent greenhouse gas. Within the Landfill Directive the UK has three targets to meet, measured as a percentage of the tonnage of BMW generated in 1995 ('the 1995 baseline'). These require the tonnage of BMW to landfill to be:

⇨ No greater than 75% of the 1995 baseline by 2010

⇨ No greater than 50% of the 1995 baseline by 2013

⇨ No greater than 35% of the 1995 baseline by 2020.

Recovery rate from non-hazardous construction and demolition waste

UK estimates of recovery rates from non-hazardous construction and demolition (C&D) waste have been calculated for reporting against the EC Waste Framework Directive. Accurately quantifying C&D waste is challenging and whilst the absolute tonnage figures are subject to a relatively high level of uncertainty, there is not a significant impact on the final recovery rate. Under this Directive there is a target for the UK to recover at least 70 per cent of non-hazardous C&D waste by 2020, which it is currently meeting.

Packaging waste

UK estimates of recovery/recycling rates for packaging materials have been calculated for reporting against material-specific targets set by the EC Directive 94/62/EC on packaging and packaging waste. The Packaging and Packaging Waste Directive (as amended) set minimum recovery targets (60 per cent) and recycling targets (55 per cent) for packaging waste, to be met by 31 December 2008, as well as material-specific recycling targets. These are 60 per cent for glass, 60 per cent for paper and board, 50 per cent for metals, 22.5 per cent for plastics, and 15 per cent for wood. Since 2008, Member States must continue to meet these minimum targets, but they have the freedom to set higher domestic targets if they so choose.

15 December 2016

⇨ The above information is reprinted with kind permission from the Department for Environment Food & Rural Affairs. Visit GOV.UK for further information.

Why are some European cities better than others at dealing with garbage?

An article from The Conversation.

THE CONVERSATION

By Ian Williams, Professor of Engineering and the Environment , University of Southampton

The end of Beirut's eight-month garbage crisis may be in sight, now that trucks have begun removing the piles of rubbish accumulating in the streets to temporary landfill sites. The problems began when the city's main landfill site was closed last July, after local residents protested that it was at capacity. No alternative sites were provided for waste disposal, so garbage began to pile up on the city's peripheries, and eventually in the city itself.

When crises like this occur in other countries, Europeans tend to congratulate themselves on their efficient waste management systems. But the recent strike by garbage collectors in the Spanish city of Málaga – which resulted in enormous rubbish heaps choking the streets – shows how not all EU members manage their waste well.

It's not a new situation in Europe either: the Campania region in southern Italy has suffered serious problems with municipal waste management since the mid-1990s. The area between Naples and Caserta has even been nicknamed the "land of fires", as a consequence of the frequent blazes burning up huge mounds of illegal and hazardous wastes.

At one point in 2007, the US Embassy in Rome warned Americans against travelling to Naples and its surrounds, citing health risks. The city's tourist trade still suffers from its reputation for dirty streets.

So why do some European countries manage their waste well, while others lurch from catastrophe to catastrophe? After all, the EU has issued directives for all its member states regarding waste disposal, so you'd expect some consistency.

Not all waste is equal

For one thing, the amount of municipal waste generated differs substantially between EU countries, ranging from 747kg per capita in Denmark, to 272kg per capita in Romania in 2013. These variations reflect differences in economic wealth and consumption patterns, but also depend on how municipal waste is collected and managed.

Landfill has generally been the main method of municipal waste treatment and disposal in the EU. But a defining moment occurred in 1975, when EU laws introduced the so-called "waste hierarchy".

The waste hierarchy indicates an order of preference for different measures to reduce and manage waste. The hierarchy prioritises waste prevention: this could include using environmentally-friendly product designs, implementing local waste prevention plans and financial disincentives such as landfill tax. When waste is created, the hierarchy then gives priority to direct re-use, then to recycling and recovery methods – such as energy recovery from waste to produce heat or electricity – and last of all, to disposal.

It was optional for member states to adopt the waste hierarchy; but there was an expectation that it would be included within national waste management laws. It was hoped

Municipal waste generated by country in selected years (kg per capita), 1995–2015						
	1995	2000	2005	2010	2015	Change (%)
Belgium	455	471	482	456	419	0.8
Denmark	521	664	736		789	51.4
Germany	623	642	565	602	625	0.3
Ireland	512	599	731	624		
Greece		412	442	532		
Spain	505	653	588	510	434	-14.1
France	475	514	530	533	502	5.7
Croatia		262	336	3379	393	
Italy	454	509	546	547	486	7.0
Cyprus	595	628	688	689	638	7.2
Poland	285	320	319	316	286	0.4
Portugal	352	457	452	516		
Finland	413	502	478	470	500	21.1
Sweden	386	428	477	439	447	15.8
United Kingdom	498	577	581	509	485	-2.6

Source: eurostat, Statistics Explained

that the hierarchy would protect the environment, conserve resources and minimise the amount of waste generated.

Slow on the uptake

For a couple of decades, very few countries took any notice of the hierarchy. High levels of landfilling continued across the EU throughout the 1990s. It wasn't until 1999 that addressing the issue became a political priority, which led to the EU's landmark landfill directive.

Specific targets were not set for the overall reduction of disposal via landfill, but for the amount of plant and food waste sent to landfill. Further directives emphasised the need for member states to adopt the waste hierarchy. The most significant was the EU Waste Framework Directive, introduced in 2008, which set an objective that 50% of all municipal solid waste was to be recycled or composted by 2020.

But even these blanket targets offer no guarantee of successful waste management across Europe. For one thing, many countries break European laws without any obvious consequences; for example, Bulgaria, the Czech Republic, Poland, Romania, Slovakia and Slovenia missed the 2010 and 2013 targets for diverting biodegradable municipal waste from landfill.

What's more, every country takes a different approach: a huge range of combinations of recycling, composting, incineration and landfilling are used across the EU. Northern and central European countries have made most progress in terms of moving away from landfill: Belgium, Denmark, Germany, Austria, Sweden and The Netherlands landfill less than 3% of their municipal waste. Meanwhile, countries in the east and south have made little or no progress.

Waste not want not

The reasons for this disparity are complicated, but factors include the availability of finance, political and social will, technical skills, suitable planning and legal frameworks, and a wide range of other social, demographic, cultural and administrative factors. The notion that decisions should be taken as closely as possible to EU citizens also means that member states are obliged to come up with waste management strategies to suit their people, rather than take a uniform approach.

Meanwhile, Eastern European countries that have only recently joined the EU – such as Lithuania, Slovakia and Latvia – have not been required to use the waste hierarchy as a guiding principle. It isn't straightforward to implement the waste hierarchy within a country: new laws must be brought in, systems for data collection and monitoring set up, and separate collection and sorting systems for different kinds of waste established.

There are some good examples of what works though. Countries with strong political leadership and cross-party agreement on environmental issues – such as Germany, Denmark, Norway and Switzerland – tend to manage waste well. What's more, countries which view waste as a resource have found novel and productive uses for it.

Belgium recycles precious metals such as gold and platinum from electronics; Germany harvests biogas from the breakdown of organic materials; Wales has focused on getting people to sort waste correctly for more effective recycling and to lower greenhouse gas emissions. And the clever use of automated technology, such as underground vacuum-powered waste disposal systems, have helped keep streets clear in cities such as Barcelona, London and Copenhagen.

The protection of our environment is becoming an increasingly important global priority, and the political, commercial and health benefits of effective waste management are clear. With time, these facts should give countries the incentives they need to manage their waste effectively. Future waste management will become more about lifestyle choices and less about managing the wastes we generate.

29 March 2016

⇨ The above information is reprinted with kind permission from *The Conversation*. Please visit www.theconversation.com for further information.

© 2010–2017,
The Conversation Trust (UK)

Fly-tipping – the illegal dumping of waste

An extract from the House of Commons briefing paper

1. The problem of fly-tipping

1.1 What is fly-tipping?

Fly-tipping is the illegal disposal of household, industrial, commercial or other 'controlled' waste without a waste management licence. The waste can be liquid or solid and controlled waste includes garden refuse and larger domestic items such as fridges and mattresses.

Fly-tipping is a criminal offence pursuant to section 33 of the Environmental Protection Act 1990:

> a person shall not… deposit controlled waste, or knowingly cause or knowingly permit controlled waste to be deposited in or on any land unless a waste management licence authorising the deposit is in force and the deposit is in accordance with the licence.

There is also an associated offence relating to the unlawful deposit of waste from a motor vehicle whereby the person who controls or is in a position to control the vehicle shall be treated as knowingly causing the waste to be deposited whether or not he or she gave any instructions for this to be done.[1]

Fly-tipping is not the same as littering. More information on litter can be found in the Library Briefing Paper on *Litter: key trends, policy and legislation in England*.

1.2 What are the impacts of fly-tipping?

The Department for Environment, Food & Rural Affairs (Defra) stresses the range of impacts of fly-tipping:

> Fly-tipping is a significant blight on local environments; a source of pollution; a potential danger to public health and hazard to wildlife. It also undermines legitimate waste businesses where unscrupulous operators undercut those operating within the law.[2]

More recently, the Parliamentary Under-Secretary of State Communities and Local Government (Mr Marcus Jones) has described both litter and fly-tipping as:

> antisocial environmental crimes that pose risks to human health and animal welfare, spoil relationships between neighbours and their wider community, and affect the way people feel about the place that they call home.[3]

In 2014/15, the estimated cost of clearance of fly-tipping to local authorities in England was nearly £50 million[4] and it costs local authorities in Wales nearly £2 million a year.[5] According to a 2015 Government consultation, various sources have estimated the cost of fly-tipping on private land as between £50–£150 million a year.[6]

1.3 Why is there a problem?

The National Fly-Tipping Prevention Group (NFTPG) notes that the causes of fly-tipping are many and varied, including:

⇨ financial gain or saving by the perpetrator;

⇨ a lack of waste disposal facilities or access to them; and

⇨ laziness and an attitude that someone else will clear up the waste.[7]

The Communities and Local Government Committee highlighted some of the evidence it had received to explain the increase in fly-tipping cases in its report *Litter and fly-tipping in England*. The Government considered that the increase may, in part, be due to more incidents being reported by local authorities due to a new online reporting facility; whereas witnesses representing local councils and landowners pointed to the introduction of charges for collection of household items and garden waste by waste collection authorities.[8]

A Westminster Hall debate in February 2016 on the Communities and Local Government Committee's Report on Litter and Fly-tipping in England further discussed the problems and causes of fly-tipping. Conservative MP Bob Blackman focused on educating people in how to dispose of rubbish and highlighted the requirement to pay for some waste removal as a possible cause of fly-tipping.

> It could be one sheet that goes out when someone registers to vote, saying, "Here's advice on how you dispose of your rubbish." At a stroke, we would remove quite a few of the problems that occur with fly-tipping. From what I can see, a lot of fly-tipping is a consequence of people not getting their domestic waste collected.

> One problem with the charge is that those people will say, "Actually, I'm not prepared to pay for a service that I think should be provided by the local authority free of charge"– and has been, by the way, for a number of years–"so I'll find another way of disposing of it." Fly-tipping will become more prevalent as a result.[9]

---

1 Section 33(5), Environmental Protection Act 1990

2 Defra, Fly-tipping statistics for England, 2013 to 2014, last updated 4 February 2015
[accessed 14 September 2015]

3 HC Deb 15 February 2016: Column 187WH

4 Defra, Fly-tipping statistics for England, 2014 to 2015, last updated 20 October 2015 [accessed 5 May 2016]. See also PQ 32880 [on Fly-tipping] for estimated clearance costs since 2012/13.

5 Welsh Government, fly-tipping, last updated February 2015 [accessed 6 May 2016]

6 Defra, Waste crime: consultation on proposals to enhance enforcement powers at regulated facilities; and call for evidence on other measures to tackle waste crime and entrenched poor performance in the waste management industry, 26 February 2015 [accessed 15 September 2015]

7 NFPTG website, About fly-tipping [accessed 6 May 2016]

8 House of Commons Communities and Local Government Committee, Litter and fly-tipping in England, Seventh Report of Session 2014-15, HC 607, 14 March 2015, para 20 [accessed 15 September 2015]

9 HC Deb 25 Feb 2016, Col 179WH

Labour MP Liz McInnes highlighted the cuts to local government funding as an influencing factor:

Unfortunately, cuts to local government funding are forcing many councils to make savings by closing municipal tips, which could increase fly-tipping. The closure of the local tip in Heywood in my constituency led to great concern among residents, many of whom contacted me to share their worries that such action would exacerbate the growing problem of fly-tipping.[10]

The Parliamentary Under-Secretary of State Communities and Local Government (Mr Marcus Jones) noted in his response that:

The Government are committed to localism and the transfer of power to local communities to deal with litter and fly-tipping problems, which require a local approach tailored to the characteristics of the area and the community in which the problems occur. Like the rest of the public sector, local authorities have worked hard over the last five years, but they still need to be thinking innovatively about how they can make litter and fly-tipping-related savings while protecting existing street cleansing services and standards.[11]

2. What are the penalties for fly-tipping?

2.1 Current penalties

The penalties for fly-tipping set out in the Environmental Protection Act 1990 were increased through the Clean Neighbourhoods and Environment Act 2005. It is now a criminal offence punishable by a fine of up to £50,000 or 12 months imprisonment if convicted in a Magistrates' Court. The offence can attract an unlimited fine and up to five years imprisonment if convicted in a Crown Court.

In addition to fines, those found guilty of fly-tipping may also have to pay legal costs and compensation, which can greatly increase the financial implications of illegal dumping.

For example, in 2010 a firm was fined £95,000 for the illegal dumping of waste; in 2013 two waste criminals were fined more than £80,000 for fly- tipping asbestos waste; and in 2016 an individual was fined £47,000 for allowing waste to be illegally deposited at two different sites, causing a fly infestation as well as risk of pollution and fire.[12]

From July 2014, new sentencing guidelines produced by the Sentencing Council for England and Wales have also increased potential fines for serious environmental offences, including fly-tipping.[13]

In addition, under the Proceeds of Crime Act 2002, offenders can have assets frozen and confiscated. In 2010 two men were ordered to pay £234,393 under the provisions, the second highest confiscation achieved by the Environment Agency at the time.

There are also a range of other possible penalties including:

⇨ Fixed penalty notices: from 9 May 2016, local authorities in England can issue fixed penalty notices between £150 to £400 for small-scale fly-tipping offences pursuant to the Unauthorised Deposit of Waste (Fixed Penalties) Regulations 2016. This may be served as a criminal penalty in lieu of prosecution for a criminal offence. It is not a civil penalty (see further below).

⇨ In Scotland, fixed penalty notices can be used for any fly-tipping incidents under section 33A of the Environmental Protection Act 1990.

⇨ Seizing property: the Control of Waste (Dealing with Seized Property) (England and Wales) Regulations 2015 came into force on 6 April 2015 and apply in England and Wales. These regulations establish the procedures which a waste collection authority (most local authorities),

10 HC Deb 25 Feb 2016, Col 185WH

11 HC Deb 15 February 2016: Col 187WH

12 2 ENDS Report, Operator fined over illegal 'fly-infested' waste site, 10 March 2016
[accessed 5 May 2016] [subscription needed]

13 Sentencing Council, New sentencing guideline for environmental crimes brings
higher sentences for serious offenders, 26 February 2014 [accessed 15 September
2015]

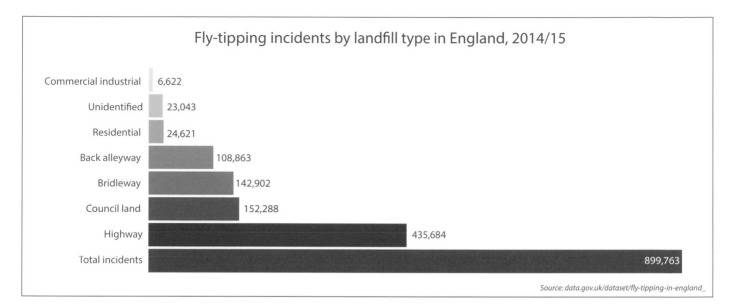

Fly-tipping incidents by landfill type in England, 2014/15

Landfill type	Incidents
Commercial industrial	6,622
Unidentified	23,043
Residential	24,621
Back alleyway	108,863
Bridleway	142,902
Council land	152,288
Highway	435,684
Total incidents	899,763

Source: data.gov.uk/dataset/fly-tipping-in-england_

the Environment Agency and Natural Resources Wales must follow once they have seized a vehicle and/or its contents because of suspected involvement concerning the transport or disposal of waste (such as fly-tipping).[14]

2.2 Introducing fixed penalty notices in England for small-scale fly-tipping

Some organisations have claimed that the penalties handed down are too low to discourage fly-tipping. A Government consultation on proposals to enhance measures to tackle waste crime ran between 26 February 2015 and 6 May 2015. The consultation sought views on enhancing enforcement powers and other measures to tackle waste crime in England and Wales. It included a call for evidence on adopting fixed penalties for small-scale fly-tipping.

In its report *Litter and fly-tipping in England* (published when the waste crime consultation was still open in March 2015) the Communities and Local Government Select Committee recommended that the Government introduce a national fixed penalty notice for small amounts of fly-tipping:

Fly-tipping is a serious problem for local authorities and private land owners, and it is increasing. There is therefore a need for local councils to increase their efforts both to deter fly-tipping and to penalise those who engage in it. We accept that prosecution is often difficult and costly and as a result the number of convictions for fly-tipping is low. The Government should introduce a national fixed penalty notice for small amounts of fly-tipping....[15]

The Conservative Party Manifesto 2015 contained a commitment to allow councils to tackle small-scale fly-tipping through fixed penalty notices. Following this, the Conservative Government published its response to the waste crime consultation in October 2015 in which it confirmed that Defra would introduce legislation at the earliest opportunity to introduce such fixed penalty notices:

The Government welcomes the high level of support for the introduction of FPNs for fly-tipping. We agree that they should only be used for small scale fly-tipping and enforcement bodies should continue to bring prosecutions in serious cases.[16]

The Government consultation response also explained that "given the high level of support in the Call for Evidence, the Welsh Government will also be exploring this further in Wales".[17]

The Government response to the Communities and Local Government Select Committee report was published in December 2015 and confirmed that, in accordance with its response to its waste crime consultation, it planned to give local councils the power to tackle small scale fly-tipping through penalty notices as an alternative to prosecutions.[18]

From 9 May 2016, local authorities in England can issue fixed penalty notices between £150 to £400 for small-scale fly-tipping offences pursuant to the Unauthorised Deposit of Waste (Fixed Penalties) Regulations 2016.

9 May 2016

⇨ The above information is reprinted with kind permission from The House of Commons Library. Please visit www.parliament.uk for further information.

© Crown copyright 2017

14 The powers for seizure arise under either section 5 of the Control of Pollution (Amendment) Act 1989 or section 34B of the Environmental Protection Act 1990

15 House of Commons Communities and Local Government Select Committee, Litter and fly-tipping in England, Seventh Report of Session 2014-15, HC 607, 14 March 2015, para 52 [accessed 15 September 2015]

16 Defra, Government response to consultation on enhanced enforcement powers and other measures to tackle waste crime and entrenched poor performance in the waste management industry, October 2015, p29 [accessed 5 May 2016]

17 Ibid

18 Government response to the Communities and Local Government Select Committee's Seventh Report of Session 2014-15 on Litter and Fly-tipping in England, Cm 9097, December 2015, para 27 [accessed 5 May 2016]

WEEE (e-waste) around the world

Many developing nations throughout the world lack regulations for disposing and recycling of electronic waste that are sufficient to protect workers, public health and ecosystems surrounding these recycling facilities. Repair and reuse are popular, and large informal recycling often occurs in developing countries; but in both developed and developing nations, landfilling of e-waste is a practice that leads to public health and environmental issues. Electronic and electrical devices account for a growing percentage of the global waste stream as consumption of new devices increases yearly.

China

Sources and volume

In 2014, over 40 million tonnes of electronic waste were generated around the world, up to 70% of which was discarded in China. China is the world's largest importer of electronic goods, and the world's largest importer of WEEE (e-waste). While importing e-waste was formally banned in 2000, China remains the largest e-waste dumping site in the world, receiving e-waste from the US, Europe, South Korea and Japan. Despite the ban, e-waste still arrives in China

through illegal means, mixed shipments, and imports through neighbouring areas such as Hong Kong where the mainland China ban on e-waste imports does not apply. Hong Kong borders Guangdong Province, where Guiyu – China's largest recycling centre – is located. Over 700,000 people are employed in the WEEE (e-waste) industry in China, 98% of whom worked in the informal recycling sector, where recycling practices are not sufficiently regulated to protect workers and environment.

Collection and recycling

China does not have a municipal e-waste collection network system in place, but has some company-sponsored takeback programmes. Most of these are in urban areas, and as rural areas experience increased economic prosperity, means to recycle e-waste in non-urban areas will need to be addressed. Informal recycling of e-waste is prevalent in China, particularly in coastal regions. Any informal e-waste recyclers not registered with the state are illegal. The informal recycling sector has developed along with improving economics and standards of living in the country, and in the 1990s, imported e-waste began to enter China as a source for cheap second-

hand goods and secondary resources. The most common outputs of informal recycling are second-hand components, refurbished appliances and materials. Informal collectors in China are often rural migrant workers who travel door-to-door in urban areas to collect e-waste and then take it to recyclers.

Anyone in China can be an e-waste collector. However, those who are authorised collectors such as retailers, chain stores and waste collection companies can participate in rebate programmes when the Government sponsors them. A 2009–2011 rebate programme for home appliances and a national e-waste legislation enacted in 2011 temporarily increased the growth of the formal sector, but without rebate programmes, the formal sector is less competitive than the informal sector. Formal recycling in China must be licensed by the provincial Environmental Protection Bureau. As of May 2012, 53 e-waste treatment enterprises in 15 provinces had this licence. Most Chinese consumers do not willingly dispose of WEEE without financial compensation: many think that such goods are still worth money, further encouraging use of informal recycling where e-waste items are paid for up front.

China has two main e-waste recycling towns which have been studied extensively – Guiyu and Taizhou. These towns typify Chinese e-waste recycling. In Guiyu, where e-waste recycling has occurred since the late 1980s, 80% of families are involved in e-waste recycling. Most of the labourers are rural migrants who take jobs of dismantling and processing e-waste for an average wage equivalent to 1.5 US$ per day. Most of these workers are women and children. A study of Chinese e-waste workers found they had significantly higher levels of PBDEs, lead and other toxins in their bodies, leading to adverse health effects and interference with thyroid and other hormone systems.

India

Sources and volume

India is the second largest e-waste generator in Asia, and more than 90% of the domestically generated e-waste ends up in the informal recycling sector. Almost

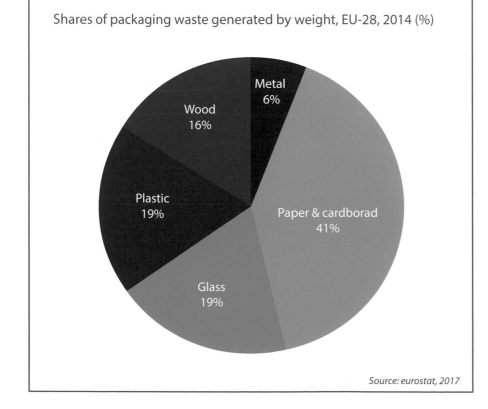

Shares of packaging waste generated by weight, EU-28, 2014 (%)

Metal 6%

Wood 16%

Plastic 19%

Paper & cardborad 41%

Glass 19%

Source: eurostat, 2017

70% of e-waste generated domestically in India is from the Government, public and private industrial sectors, while individual households account for only 15% of the total.

70% of the e-waste that is processed at recycling units in New Delhi are exported or dumped by developed countries. Delhi is the leading processing center of e-waste in the country: about 25,000 workers illegally refurbish up to 20,000 tonnes of e-waste annually. India generates about 3.5 million tonnes of e-waste every year, and imports another 50,000 tonnes from countries including the US, Canada, EU, Australia and countries in the Middle East. A 2008 law bans import of hazardous waste, including e-waste, for disposal into India. However, import of waste is permitted for reuse, recycling and reprocessing. E-waste still gets into the country through this and other significant loopholes such as policies that allow donation imports and porous ports with lack of enforcement capabilities.

Collection and recycling

India's informal recycling flourishes in Bangalore, Chennai, Delhi and New Delhi. An estimated 25,000 workers are employed at unregulated e-waste scrap yards in Delhi alone. India has a lucrative refurbishing market fuelled by cheap migrant labour. Informal recycling centers are usually located in urban slums. Many workers work out of their homes, exposing themselves and their families to toxins. Child labour is often used in this sector. A large source of domestic e-waste supplied to these recycling centres is collected by waste pickers who buy scrap electronics from households. Most imported e-waste goes to the informal recycling sector, and waste that is imported from developed countries that processors choose not to refurbish or dismantle often gets dumped into rivers.

In 2009, formal recyclers came together to form an association, but they face stiff competition from the informal sector and capture only about 10% of the total share of the e-waste market. The formal sector lacks sufficient collection and disposal systems and appropriate technologies, making it difficult for households to transfer WEEE to them. Formal recyclers also have licences to import e-waste from abroad. Most WEEE items from households are refurbished and then resold – very few obsolete electronics actually end up in e-waste recycling and processing streams.

Ghana
Sources and volume

Electronic and electrical goods in Ghana are imported mainly from Europe, North America, and Asia. Electronics imports into Ghana in 2009 were approximately 215,000 tons, with 30% being new products and 70% second hand or used goods. Many of the used items imported are labelled donations. The second-hand market in Ghana is large: most obsolete items go to repair stores instead of to municipal waste. Out of all the electronics in Ghana that were at the end of their usable life in 2009, 57% went to repair, 34% to recycling, 8% to storage and only 1% was collected in municipal waste.

Collection and recycling

Ghana has no current regulations for recycling and disposal of e-waste and does not enforce air, water or soil quality standards. The Agbogbloshie market in Accra is the largest centre for e-waste recycling and disposal in Ghana. Small recyclers work out of sheds and in the open. Soils near the market are covered in litter and industrial oils. Informal settlements next to the site have little electricity and running water. Most work done is by manual disassembly and burning to recover wire and coils, mostly by young adults (between 21–30 years) with no or little protective equipment. Pollution occurs through soil, air and water contamination. Air pollution from burning e-waste affects not only those living near it, but also in the surrounding areas. Waste obstructs sewer systems and litter covers the ground. Children frequently come into contact with contaminated soils – a main source of toxic chemical poisoning.

Nigeria
Sources and volume

In 2010, an estimated 360,000 tonnes of e-waste were informally collected and processed in Nigeria, of which 100,000 came from illegal imports and 160,000 from private households. Over 100,000 tonnes were also collected through communal collection via household waste. Most electronics, new and used, arrives through the Lagos Seaport. Most of the used electronics arrives from Europe, but some also arrives from the US, China, Japan, Morocco, Canada and Taiwan under the guise of donations. At least 100,000 tonnes of e-waste was entering the country illegally in 2010.

Collection and recycling

There are no WEEE recycling facilities in Nigeria. Reuse activities usually revolve around refurbishing mobile phones, but batteries and accessories generate a large contribution to landfills. Most crude metal dismantling is performed in Lagos, and from there is often exported to China or other Asian countries.

The informal sector is the major collector of e-waste, with waste pickers collecting from households or picking through dumps. This sector is dominated by young people, who then sell their scrap to others for recycling or repair. Formal recycling in the country involves pickers who scavenge through municipal dumps to remove items that still have value. About 20% of disposed items are refurbished, reused or dismantled in this way. Municipal solid waste facilities collect and burn all other waste before disposal into landfills. Most household waste ends up in open dumpsites and landfills after burning, leading to leaching of chemicals into soil and groundwater.

United States of America
Sources and volume

In the US in 2010, 384 million units of consumer electronics reached the end of their life and were disposed. The US generated over 3.4 million tons of e-waste in 2012; 29% of this was recycled, while the remaining amount went to incinerators or landfills. Mobile devices were the most disposed-of item, but together the 141 million units at the end of their life in 2009 made up less than 1% of the total weight of all discarded items.

Collection and recycling

A number of manufacturers and retailers participate in takeback programmes. Legislation in 25 states exists to cover the disposal and collection of electronic waste: most of this legislation follows two different models. The first model – enacted by 24 states and Washington D.C. – encourages extended producer responsibility, where the manufacturer pays to collect and recycle their products. What types of products are covered under this law varies between states. The second type of program is the advanced recycling fee model, where consumers

pay retailers a fee at the time of purchase, which is deposited into a statewide recycling fund.

In 2011, a taskforce released the National Strategy for Electronics Stewardship, which provides a framework to encourage electronics stewardship in the US. This established a best practices guideline for the Federal Government to follow, and voluntarily for retailers and manufacturers to adhere to. The NSES also sought to certify electronics recycling and refurbishment facilities: it is currently a voluntary process, but the NSES is a standard against which third-party auditors assess their practices. By June 2014, 565 facilities obtained voluntary certification.

European Union

Sources and volume

EU countries dispose of an estimated 6.5 million tonnes of WEEE yearly. Almost half of this volume is large household appliances. An estimated one million tonnes of WEEE is thrown away by UK households yearly. In Germany, an estimated 1–1.4 million tonnes of WEEE is discarded annually from private households.

Collection and recycling

The EU WEEE Directive requires manufacturers and importers to take back products from consumers and ensure they are disposed of using environmentally sound methods. However, the European Commission reports that only one third of e-waste gets collected, treated and reported according to the directive.

Despite formal WEEE collection in Germany, some consumers still dispose of their unwanted WEEE with regular household waste. Gaps in laws allow large amounts of e-waste declared for recycling to be shipped to countries like India, China and Nigeria. Public waste authorities collect discarded products, while retailers may also voluntarily accept returned WEEE and transport it to the producer or to waste authorities. In the UK, producers are required by law to join producer compliance schemes and must finance the treatment, recovery, recycling and environmentally sound disposal of WEEE. Handling is done by a network of authorised treatment facilities and approved exporters. Distributors are also required to facilitate free takeback through offering in-store takeback schemes

Sources

⇨ Amoyaw-Osei, Y., Agyekum, O.O., Pwamang, J.A., Mueller, E., Fasko, R., & Schluep, M. (2011). Ghana e-Waste Country Assessment: SBC e-Waste Africa Project: E-waste Africa Project of the Secretariat of the Basel Convention. Swiss Federal Laboratories for Materials Science and Technology: Switzerland.

⇨ Caravanos, J., Clark, E., Fuller, R., & Lambertson, C. (2011). Assessing Worker and Environmental Chemical Exposure Risks at an e-Waste Recycling and Disposal Site in Accra, Ghana. Journal of Health and Pollution. 1, 16-25.

⇨ Electronics Take Back Coalition (2014). Facts and Figures on E-Waste and Recycling. Retrieved from: www.electronicstakeback.com

⇨ Interagency Task Force on Electronics Stewardship (2014) Moving Sustainable Electronics Forward: An update to the National Strategy for Electronics Stewardship.

⇨ Ongondo, F.O., Williams, I.D., & Cherrett T.J. (2011). How are WEEE doing? A global review of the management of electrical and electronic wastes. Waste Management. 31, 714-730.

⇨ Ogungbuyi, O., Nnorom, I.C., Osibanjo, O., & Schulep, M. (2012). E-waste Country Assessment Nigeria: E-Waste Africa project of the Secretariat of the Basel Convention. Swiss Federal Laboratories for Materials Science and Technology: Switzerland.

⇨ Perkins, D.N., Drisse, M.-N.B., Nxele, T., & Sly, P. (2014). E-Waste: A Global Hazard. Annals of Global Health. 80(4), 286-295.

⇨ Rajya Sabha Secretariat Research Unit (2011). E-Waste in India. Rajya Sabha Secretariat Research Unit Occasional Papers: New Delhi.

⇨ Schultz, J., & Oleen, B. (2015). Electronic Waste Recycling. National Conference of State Legislatures. Retrieved from: http://www.ncsl.org/research/environment-and-natural-resources/e-waste-recycling-legislation.aspx

⇨ United Nations University (2014). Global E-Waste Volume Hits New Peak in 2014: UNU Report. Retrieved from: http://unu.edu/news/news/ewaste-2014-unu-report.html

⇨ U.S. EPA (2011). Electronics Waste Management in the United States through 2009: Executive Summary.

⇨ Wang, F., Kuehr, R., Ahlquist, D., & Li, J. (2013). E-Waste in China: A country Report. United Nations University: Institute for Sustainability and Peace.

⇨ The above information is reprinted with kind permission from The University of Washington and Coming Alongside Environmental Services. Visit ewaste.ee.washington.edu/students/locations/ or www.comingalongside.org/Ewaste/ for further information.

How much waste is really produced at UK music festivals?

How many skips does festival waste fill? How many tonnes are produced? Is it recycled? Do our beloved UK music festivals have initiatives in place to improve on their environmental impact? We have investigated to bring you all the facts and figures.

People attend music festivals to have a good time, and perhaps escape everyday life for a few days, but have you ever stopped to think about how much waste is produced by these festivals – some of which are so big they are essentially a large temporary town? Have you ever wondered about how much of the waste produced is recycled, or what green initiatives they have in place to cut waste and litter? With the number of music festivals being hosted in the UK growing by 34 per cent in just four years, these are questions that need to be asked.

How are festivals trying to reduce and manage their waste?

Our research has led to us uncovering a vast array of interesting initiatives at music festivals, most of whom are trying hard to improve on their green statistics from previous years, and lessen their environmental impact. We go into detail about a few of these below.

General waste/rubbish

Festivals are very keen to produce the least amount of waste possible, as they know they will suffer criticism if not. It has been found that half of festival-goers would happily pay more to attend a greener event, as opposed to an event known for its litter and waste problems. Also, most say "noise, waste and traffic have a negative impact on the festival experience".

Some festivals – including Acoustic Festival and Larmer Tree Festival – recycle ALL of the waste produced onsite, meaning the waste which is not recyclable in the classic sense gets turned into energy.

Cambridge Folk Festival recycled all but 2% of their waste in 2015, and are aiming for 100% recycling this year.

They're also hoping to recycle at least 50% onsite this year, as last year 36% of the sorting and recycling took place offsite, which uses more energy.

In 2014, Glastonbury Festival recycled 54% of the waste produced onsite.

Reading Festival only managed to recycle about a quarter of their total waste in 2013 and 2014. In fact, there was a 205-tonne increase in waste produced on site between 2013 and 2014, combined with a 2% drop overall in recycling. 2013 saw 433 tonnes of waste and 169 tonnes recycled, and 2014 saw 596 tonnes and 214 tonnes recycled.

However, their sister festival, Leeds Festival kicked Reading's butt in 2014 and managed to recycle a healthy 82.4% of their waste! There was 13 times more waste recycled at Leeds Festival in 2014 than in 2013. 887 tonnes of waste were sent to landfill in 2013, with 54 tonnes being recycled. Then, in 2014 the tables turned dramatically, and 120 tonnes of waste was sent to landfill with 568 tonnes recycled. That is still 120 tonnes too much being sent to landfill, but certainly a step in the right direction.

Litter

Some festivals issue rubbish bags to festival-goers on arrival, and add a charge to their tickets, which is refunded if a bag of rubbish is returned at some point during the festival. This aims to keep litter to a minimum, and keep rubbish off the floor, where it can quickly impact on the local environment.

Cup deposit schemes are becoming increasingly popular and successful at festivals, where you only get your deposit back if you return the reusable cup you've been using. Stepping Stones Festival told us this works really well for them, with many people taking the cups home as a cheap souvenir, and the majority of the rest being returned for the deposit. This way, none end up on the floor or in landfill.

All festivals seem to have increased the number of bins on site, which are clearly labelled with what they are for (general waste, plastic, paper, etc.). The problem seems to be bins overflowing with rubbish, as they are not emptied often enough.

Plastics

Plastic is often seen as the most evil of the recyclables – mainly due to the toxins it can contain, which can leach into the environment. It's also not as easy to recycle (and therefore uses more energy) than other materials. For these reasons, some festivals have cut plastic out of their disposable food and drink packaging altogether. Limetree Festival is this year going plastic-free, and will be providing free tap water throughout the festival, as well as encouraging festival-goers to take advantage of the natural spring onsite. As mentioned earlier, Glastonbury are this year giving out stainless steel drinks containers for a £5 deposit, in an attempt to cut their plastic bottle consumption: around one million plastic bottles are usually used and disposed of at Glastonbury each year!

Tents

Tents are a modern nightmare for the organisers of music festivals, thanks to increasingly low tent prices, poor tent quality and apathetic festival-goers. Some festivals, including Bestival, collect up the abandoned tents and donate them to humanitarian charities – recycling them for others to use.

Interestingly, the Isle of Wight festival now features special 'Love Your Tent' camping fields, in which everyone signs a code of conduct when they arrive, agreeing to remove all of their belongings when they leave. And it works! In 2014, 1,450 campers stayed in the designated fields, and not one tent or item or rubbish was left behind! Compare that to your typical music festival camping aftermath, and you realise they might be on to something there.

Toilets

Everyone's least favourite aspect of music festivals: the toilets. In the past, and to this day at some festivals, all of

the toilets are chemical – meaning they are far from environmentally friendly, and are adding to landfill. However, some festivals are moving towards better options, with Glastonbury banishing chemical loos forever (save 150) in favour of eco-friendly long-drop toilets and compost loos, the results of which are later spread on farms to help crops grow.

Glastonbury Festival has just been fined in court for causing pollution during last year's festivities, due to a human sewage leak, which ran into the nearby waterways and killed over 40 fish – some of which were very rare. Perhaps the move to long drop and compost loos will stop any repeats of this incident.

Are attempts to be more environmentally friendly working?

Some festivals are doing extraordinarily well at keeping waste to a minimum, minimising their environmental impact, and leaving no trace after the event. Others, not so well. Take Latitude Festival for example: this festival created 157 tonnes of landfill waste in 2014 and 43 tonnes of recycled waste. The most worrying thing about these figures is, Latitide have a green policy in place which involves handing out clear bags to all party-goers for their recyclables, a cup deposit scheme,

and entry to a prize draw for those returning a bag of waste to their waste points. Whilst in 2014 they did improve on 2013's efforts, more needs to be done.

What's the key to environmental success?

Aside from cancelling all our music festivals, which isn't going to happen, how can we make them as environmentally friendly as possible, cut waste production and cut the carbon footprint? Mike of Acoustic Festival, which runs 75% of the festival on solar energy and recycles all their waste, says their environmental success is down to: "attract[ing] a green and tidy ethos/person who all do their bit".

Some green initiatives which, according to our research, seem to work well at reducing waste and litter include:

⇨ Getting attendees to read and sign a green agreement

⇨ Banning 'disposable' plastic from the site

⇨ Giving every festival-goer a reusable cup, using a deposit scheme

⇨ Ensuring all catering vendors use compostable, biodegradable, or at least easily recyclable containers and wrappers.

⇨ Issuing all festival-goers with bags for their rubbish/recycling, and a

reward scheme for returning the bags full.

Of course, there are many other environmental aspects to take into account with music festivals, including carbon footprints, and the lasting impact on the land used.

However, it is great that the green impact of music festivals is being thought about nowadays – for so long, it was ignored by most festivals. The reduction of waste production is the key, as opposed to just recycling as much as possible: prevention, as they say, is so much better than a 'cure'. I look forward to seeing what new initiatives are trialled in the future, and hope to see a real focus on waste reduction.

10 June 2016

⇨ The above information is reprinted with kind permission from Forge Skip Hire. Please visit www. forgeskiphire.co.uk for further information.

⇨ The original blog can be found at: https://www.forgeskiphire.co.uk/blog/music-festival-waste/

First evidence of deep-sea animals ingesting microplastics

Scientists have found evidence of microfibres ingested by deep sea animals, revealing for the first time the environmental fallout of microplastic pollution.

The UK Government recently announced that it is to ban plastic microbeads, commonly found in cosmetics and cleaning materials, by the end of 2017. This followed reports by the House of Commons Environmental Audit Committee about the environmental damage caused by microbeads. The Committee found that a single shower can result in 100,000 plastic particles entering the ocean.

Researchers from the universities of Bristol and Oxford, working on the Royal Research Ship (RRS) *James Cook* in the mid-Atlantic and south-west Indian Ocean, have now found evidence of microbeads inside hermit crabs, squat lobsters and sea cucumbers, at depths of between 300m and 1,800m. This is the first time microplastics – which can enter the sea via the washing of clothes made from synthetic fabrics or from fishing line nets – have been shown to have been ingested by animals at such depth.

The results are published in the journal *Scientific Reports*.

Laura Robinson, Professor of Geochemistry in Bristol's School of Earth Sciences, said: "This result astonished me and is a real reminder that plastic pollution has truly reached the furthest ends of the Earth."

Microplastics are generally defined as particles under 5mm in length and include the microfibres analysed in this study and the microbeads used in cosmetics that will be the subject of the forthcoming government ban.

Among the plastics found inside deep-sea animals in this research were polyester, nylon and acrylic. Microplastics are roughly the same size as 'marine snow' – the shower of organic material that falls from upper waters to the deep ocean and which many deep-sea creatures feed on.

Dr Michelle Taylor of Oxford University's Department of Zoology, and lead author of the study, said: "The main purpose of this research expedition was to collect microplastics from sediments in the deep ocean – and we found lots of them. Given that animals interact with this sediment, such as living on it or eating it, we decided to look inside them to see if there was any evidence of ingestion. What's particularly alarming is that these microplastics weren't found in coastal areas but in the deep ocean, thousands of miles away from land-based sources of pollution."

The animals were collected using a remotely operated underwater vehicle. The study, funded by the European Research Council (ERC) and the Natural Environment Research Council (NERC), was a collaboration between the University of Oxford, the University of Bristol, the Natural History Museum in London, and Staffordshire University's Department of Forensic and Crime Science, which made sure the results were robust and the study was free from potential contamination.

Dr Claire Gwinnett, Associate Professor in Forensic and Crime Science at Staffordshire University, said: "Existing forensic approaches for the examination of fibres are tried and tested for their robustness and must stand up to the scrutiny of the courts of law. These techniques were employed in this research in order to effectively reduce and monitor contamination and therefore provide confidence in the fact that the microplastics found were ingested, and not from the laboratory or other external contaminant.

"Using forensic laboratory techniques, we have identified that microplastics are present in ingested material from deep sea creatures. Forensic science is still a fairly new science, but we are delighted that our work and techniques are starting to inform other sciences and important environmental research such as this."

30 September 2016

⇨ The above information is reprinted with kind permission from the University of Bristol. Visit www.bristol.ac.uk for further information.

'Missing plastic' in the oceans can be found below the surface

An article from The Conversation.

By Julia Reisser, Oceanographer, PhD candidate researching plastic pollution and sea turtles, University of Western Australia, Charitha Pattiaratchi, Professor of Coastal Oceanography, University of Western Australia and Maira Proietti, Adjunct Professor, Fundação Universidade Federal do Rio Grande

The world's ocean contains trillions of plastic fragments coming from packaging, fishing gear and other synthetic objects that break down at sea over time.

Most of what is known about these ocean plastics comes from surface net sampling, where the top 15cm of water is filtered to collect particles larger than 0.3mm.

Now we have published the first ever high-resolution depth profiles of ocean plastics in the journal *Biogeosciences* and data repository Figshare.

Most of the submerged plastics were very small – less than 1-mm across. Previous studies had noticed that tiny plastics were missing from the oceans.

We show that at least a fraction of this missing plastic is still adrift at sea, but at depths greater than the surface layer that is usually sampled by scientists.

A better characterisation of the vertical distribution of marine plastic pollution will improve predictions of plastic loads, particle sizes, and ecological impacts in the world's oceans.

Vertical distribution of buoyant plastics

Buoyant plastics can be pushed underwater by the turbulence created by wind and waves. Models predict that the number of plastic particles decreases abruptly over the first few metres of the water column. However, until now, no subsurface measurements existed to test this prediction.

We developed a new device that collects samples from the sea surface at 50cm intervals, down to a depth of five metres. This device was used to sample one of the world's major plastic pollution hotspots: the North Atlantic "garbage patch".

Buoyant plastics were concentrated at the sea surface, with both numerical and mass concentrations decreasing exponentially with water depth. Nevertheless, under stronger winds this decrease was less abrupt. Our results match relatively well with those predicted by scientific models.

Look below the sea surface

The speed in which buoyant plastics return to the sea surface after being pushed into deeper waters by turbulence is an important parameter for predicting the depth profiles of marine plastic pollution. We found that smaller plastics present lower rising speeds, being therefore more susceptible to transport into deeper layers.

Even under light wind conditions, many of the tiniest plastics were still hidden underwater. This indicates that previous studies using surface-only samples are biased towards larger plastic pieces.

Two major studies last year headed by ocean scientists Andres Cózar and Marcus Eriksen both concluded that there are major losses of small plastics from oceans. We show that at least a fraction of this "missing" plastic is just under the sea surface.

More at-sea and experimental work is required to further quantify this effect and develop models capable of estimating depth-integrated size distribution of buoyant plastics drifting at sea.

What's next?

Samples of ocean plastics from below the surface are still very scarce. Further multi-level sampling is extremely important to help us estimate how much plastic is actually in our oceans, and understand its ecological impacts.

Knowing how deep plastics go will help determine the chance of animals inhabiting different depths to encounter and interact with plastic items. For instance, sea birds, turtles, and mammals, which breathe air and use the sea surface for daily activities, present high rates of plastic ingestion and entanglement.

Ocean plastics are also enhancing ocean drift opportunities and causing damage to biota and habitats. Plastic items harbour organisms such as microbes, invertebrates and fish, which can disperse across oceans and potentially invade non-native habitats.

We emphasise that estimates on the amounts and impacts of plastic pollution across the oceans, both in coastal and oceanic environments, are urgently required. Such research is crucial to better inform those aiming to reduce the flow of plastics entering this environment, and develop mitigation strategies for this worldwide problem.

26 February 2017

⇨ The above information is reprinted with kind permission from *The Conversation*. Please visit www.theconversation.com for further information.

From L'Oréal to Revlon, which brands are polluting the ocean with microbeads?

After discovering that 170 types of seafood contained traces of microplastics, Greenpeace East Asia decided to put 30 of the world's biggest cosmetic and personal care brands to the test.

By Taehyun Park

Microbeads might be best known to you as the coloured beads in your shower gel and facial scrubs, but more recently they've gained notoriety for wreaking havoc on our ecosystems and marine life.

The tiny pieces of plastic are added to everyday cosmetic products as an exfoliating agent, or for colour and texture. They're tiny enough to travel down your plughole and right through water filtration systems and into our rivers, lakes and oceans.

A single tube of face wash can contain up to 360,000 of these tiny plastic spheres. That means that once we have finished washing our faces or brushing our teeth, we unwittingly release thousands of pieces of plastic into our environment, where they go on to 'gently exfoliate' the digestive tracts of seabirds and even enter the food chain.

They can also act as agents to absorb and release toxic chemicals around the sea and into the marine life that ingests them.

How do the companies stack up?

Greenpeace East Asia decided to put 30 of the world's top companies to the test and rank them according to the strength of their commitment to getting rid of microbeads once and for all.

While most of the brands claim to have their plastic pollution problem under control, not one of them succeeded in meeting Greenpeace's environmental standards, meaning that they still have the potential to allow this contamination into our waterways.

Even the top scoring brands in the rank, like Beiersdorf, which has allegedly fulfilled all its commitments to its microbead ban pledge, have only taken action to remove one type of plastic – polythene – from its products, which gives a free rein to other polluting plastics.

The harmful effects of microbeads are now well known and we're gathering more and more evidence that they're bad news. Many of the world's biggest brands have made pledges to rein in these toxic terrors, but they're still not that simple for consumers to avoid.

Firstly, while some companies proudly tout the presence of 'skin-polishing' microbeads in their product descriptions, others contain microbeads that can be barely seen with the naked eye and only appear in the ingredients list as polyethylene, polypropylene or polystyrene.

Secondly, many brands have made promises to do the right thing and ban the beads, but each brand has its own, narrow or confusing definition of what constitutes a microbead. These definitions can vary from function of the product, role of the microbead and can even change the shape of the microbead, creating loopholes that could allow the inclusion of microbeads that don't fit into these limited definitions.

What's the solution?

So how to get these pesky microbeads out of our products and out of our oceans? The solution is simple. Our governments need to step in and enforce a total ban on the sale and production of all solid microplastic ingredients in all personal care products.

The good news is, it's already happening. The US announced a ban on microbeads in January of this year, while campaigns to do the same are building momentum all around the world.

In the meantime, you can vote with your wallet and choose brands that don't add to this pointless pollution. Check out Fauna & Flora International's *Good Scrub Guide* or download the Beat the Microbead app and send a clear message to manufacturers that microbeads are unwanted and unnecessary.

Taehyun Park is an Oceans Campaigner at Greenpeace East Asia.

Greenpeace East Asia consulted Fauna & Flora International on expected good practice with respect to corporate commitments to ending microplastic ingredient use.

21 July 2016

Information was correct at the time of writing.

The original article can be found at: http://www.greenpeace.org/international/en/news/Blogs/makingwaves/microbeads-microplastics-ranking-personal-care-cosmetics/blog/57082/

⇨ The above information is reprinted with kind permission from Greenpeace. Please visit www.greenpeace.org for further information.

© Greenpeace 2017

What is marine litter and why it is of concern?

Marine litter (or debris)* is waste created by humans that has been discharged into the coastal or marine environment. It is defined as "any anthropogenic, manufactured, or processed solid material (regardless of size) discarded, disposed of, or abandoned in the environment, including all materials discarded into the sea, on the shore, or brought indirectly to the sea by rivers, sewage, storm water, waves, or winds"

Just as human activities are varied and widespread, so are the sources of litter. The sources may be located directly at sea, on the coast or further inland. Litter can be transported over long distances and into all marine habitats – from the surf zone all the way to remote mid-oceanic gyres and the deep sea floor. Like other pollutants, marine litter affects habitats, ecological function and the health of organisms of the ecosystems where it accumulates.

Any human-made object that does not naturally degrade within days or months can potentially become marine litter if it is not properly managed. Common litter items are made of paper, wood, textiles, metal, glass, ceramics, rubber and plastic discarded by humans.

Between 60 and 90 per cent – sometimes as much as 100 per cent – of the litter that accumulates on shorelines, the sea surface and the sea floor is made up of one or a combination of different plastic polymers. The most common items, constituting over 80 per cent of the litter stranded on beaches are cigarette butts, bags, remains of fishing gear, and food and beverage containers. Likewise, 90 per cent of the litter collected from sea floor trawls is made up of plastic.

Plastics have only been mass-produced for around 60 years and therefore it is impossible to know with certainty how long they last in the marine environment. Most types of plastic are not biodegradable (Andrady 1994). In fact, they are extremely durable.

This means the majority of polymers manufactured today will persist for decades and probably for centuries, if not millennia. So-called degradable plastics may persist for a long time because their degradation depends on physical factors, such as exposure to light, oxygen and temperature. Biodegradable plastics also decompose through the mediation of certain micro-organisms. Plastics labelled as biodegradable, designed to undergo certain degrees of degradation in landfills or in terrestrial environments, may still persist for long periods under marine conditions. Full degradation of a plastic item implies complete breakdown and decomposition into water, carbon dioxide, methane and other non-synthetic molecules. For the large majority of plastic items, even if they disintegrate by breaking down into smaller and smaller plastic debris under the influence of weathering, the polymer itself may not necessarily fully degrade into natural chemical compounds or chemical elements under marine conditions.

In addition to polymers, additives such as flame retardants (e.g. polybrominated diphenyl ethers), and plasticisers (e.g. phthalates) are also mixed into synthetic materials to increase their flexibility, transparency, durability, and longevity. Some of these substances, present in most plastic objects found in the marine environment, are known to be toxic to marine organisms and to humans.

The plastic used in the manufacture of an object depends on its intended use. The type of plastic will determine the ease with which an object can be recycled. Some plastics cannot be recycled, which means they enter the waste management system. If they make it into the marine environment, plastics that are less dense than sea water will float at the surface. Floating objects can be readily transported by wind, waves and surface currents and become widely dispersed across the ocean. Plastics that are denser than sea water will sink to the sea floor and accumulate or be redistributed, along with other sedimentary particles, through bottom sedimentary processes.

Marine litter comes in all sizes. Large objects may be tens of metres in length, such as pieces of wrecked vessels, lost fishing nets and lost cargo containers. Moderate-sized objects less than one metre long might include plastic bags, soda bottles or milk containers. Small spheres of expanded polystyrene are on the scale of millimetres. Micrometre-sized plastic beads are present in cosmetic products and synthetic cloth fibres or are derived from fragments broken down from larger plastic items.

There has recently been a noticeable increase in concern about the implications of pollution by small-sized debris, especially where made up of plastic. The term microplastic has been introduced to describe small plastic debris commonly less than 5mm in diameter. The concern about microplastic pollution is due to its ubiquitous presence in the marine environment. Yet it is difficult to assess its quantity because of the small size of the particles and the fact that little is known about the chemical reactions and the extent of its incorporation into the trophic chain. Investigations are also being conducted into the implications of organisms' exposure to and intake of plastic nanoparticles, particles smaller than one micron. With such limited knowledge of the ultimate ecological effects of microplastics and nanoplastics, there are concerns over their potential impacts at the level of ecosystems.

For references, see http://www.grida.no/publications/60

⇨ The above information is reprinted with kind permission from the United Nations Environment programme. Please visit www.unep.org for further information.

© United Nations Environment Programme 2017

Food waste – the issue

With the global population set to rise to over 9.5 billion by 2050 there are huge pressures facing the world's food system. We do not yet know what the future will bring for food, but we do know that business as usual is not an option.

We can all contribute to a better food future by reducing the amount of food we waste. Small changes will make a big difference, and everyone can play their part. Almost 50% of the total amount of food thrown away in the UK comes from our homes.

There are two main reasons why we throw away good food: we buy or make too much or we don't use it in time.

Businesses have a role to play in reducing food waste too. WRAP research shows that the UK manufacturing and retail sector wastes 1.9 million tonnes of food and drink a year, 1.1 million tonnes is avoidable.

Key facts

⇨ Food waste is a global issue. According to the Food and Agriculture Organization of the United Nations one third of the global food production is lost or wasted annually.

⇨ In the UK, we throw away seven million tonnes of food and drink from our homes every year, the majority of which could have been eaten.

⇨ Wasting this food costs the average household in the UK £470 a year.

⇨ Bread, fruit and vegetables are the most commonly thrown away foods.

Use-by and best-before dates

Use-by date is about safety

⇨ Use-by date is about safety and the most important date to remember! Foods can be eaten (and most can be frozen) up until the use-by date, but not after. You will see use-by dates on food that goes off quickly, such as meat products or ready-prepared salads. For the use-by date to be a valid guide, you must carefully follow storage instructions.

Best-before date is about quality

⇨ Best-before date is about quality and not safety. The food will be safe to eat after this date but may not be at its best. Its flavour and texture might not be as good. The best-before dates appear on a wide range of frozen, dried, tinned and other foods. The best-before date will only be accurate if the food is stored according to the instructions on the label.

After the use-by date, don't eat it, cook it or freeze it. The food could be unsafe to eat or drink, even if it is stored correctly and looks and smells fine.

Eat it, cook it, freeze it

⇨ Use our quick tips to save money, make time and reduce waste by eating, cooking or freezing food by its use-by date.

1. Plan your meals ahead

Try and get into the habit of checking what you already have in the fridge and freezer before you go shopping. Use up foods that are approaching their use-by date and other fresh foods like fruit and vegetables, meat, fish, cheese or milk first as these can go off over time.

2. Don't trust the sniff test!

Food can look and smell fine even after its use-by date, but that doesn't mean it's safe to eat. It could still be contaminated. You cannot see, smell or taste the bugs that cause food poisoning.

3. Batch cook and freeze multibuys

If you are taking advantage of multibuys or larger pack sizes (e.g. chicken breasts) you can freeze them individually in smaller bags. You can also cook enough for two (or more!) meals, or batch cook, and eat one and freeze some for later – this avoids waste and minimises the effort of cooking.

4. Freeze food up to the use by date

Many people believe food can only be frozen on the day of purchase; however, you can safely freeze most foods right up to the use-by date. Although it would be good to freeze it as soon as you know you aren't going to use it before its use-by date expires.

Freezing acts as a pause button. You can freeze pretty much everything. It's safe to freeze food up until the use

by date. This includes your meats and meals cooked from previously cooked and frozen meat – your roast chicken leftovers for example. Once defrosted, the pause button is off. So, just defrost your food as you need it – and eat within 24 hours.

5. Wrap up

It is best to place food in an airtight container or wrap food well in freezer bags, freezer wrap or cling film before placing in the freezer, otherwise the cold air will dry it out. Try to expel any air from freezer bags.

6. Freeze both raw and cooked meats

You can safely freeze raw and cooked meats and even cook defrosted meat into a new meal and freeze for use on another day. Simply defrost overnight in the fridge (be careful that raw meat doesn't drip on any other foods in the fridge and check it is thoroughly defrosted), use within 24 hours and cook till steaming hot.

7. Eat frozen food within three to six months

Foods can be stored in a correctly functioning freezer for a long time without going off, but it's the quality that deteriorates over time. So it's best to eat your frozen food within three to six months (but do check for any freezing instructions on the packaging). If your meat is frozen for longer – marinate it before cooking to improve texture or use herbs and spices to add flavour.

8. Defrost overnight in the fridge

You can defrost frozen food overnight in the fridge but check to make sure it is thoroughly defrosted. Once defrosted use within 24 hours and cook until steaming hot. if you need it quickly, food can be defrosted safely in the microwave, but check that there are no frozen lumps or cold spots in the middle.

⇨ The above information is reprinted with kind permission from the Food Standards Agency. Please visit www.food.gov.uk for further information.

Did Tesco waste 119 million meals last year?

In brief

Claim

Tesco wasted 119 million meals last year.

Conclusion

This is true, if you define a meal as 0.5kg of food. It's likely to equal a very small proportion of the food wasted by Tesco's customers or across its supply chain.

"Tesco food waste rose to equivalent of 119 million meals last year"

Guardian, 15 June 2016

"Supermarket threw away equivalent of 119 million meals last year despite initiatives to tackle the problem"

Daily Mail, 16 June 2016

"Tesco food waste increases to the equivalent of 119 million meals in 2016"

Independent, 16 June 2016

Tesco estimates it produced 59,400 tonnes of food waste last year across all its UK retail operations, which includes its stores, petrol stations and food depots. It's the only supermarket to publish figures on its food waste.

The figure doesn't include waste produced by Tesco's suppliers, or food bought from Tesco and wasted by households.

119 million meals

59,400 tonnes of food is the same as 119 million meals, if you define a meal as 500 grams of food and drink.

Would this look like a meal to you? On average, women in the UK aged between 19 and 65 eat about 2,500 grams of food a day, whilst men eat about 3,000 grams. Weight-wise, it's about the same as a 500ml bottle of coke.

As an aside, the different types of food wasted probably don't match the proportions you would eat them in. Nearly four-tenths of Tesco's food waste, by weight, comes from its in-store bakeries. You probably don't eat that much from that part of the store.

Defining food waste

Not all food waste is simply thrown away. For example, Tesco sells about 30% of its bakery waste to become animal feed, and the rest to companies that produce energy and fertilisers. The company says it hasn't sent any food waste directly to landfill since 2009.

Some kinds of food waste are harder to avoid than others. The Waste and Resources Action Programme (WRAP) suggests that all food waste by UK retailers is avoidable in theory, but only 44% could practically be avoided by 2025. We don't know exactly how much of Tesco's waste was avoidable.

Putting this in proportion

Last year, Tesco produced the equivalent of about 20% of the food waste produced by all UK grocery retailers in 2014, according to WRAP's estimates, whilst holding about 30% of the UK grocery market. This comparison doesn't include food sent to become animal feed because although Tesco defines it as food waste, WRAP doesn't.

If we include food sent to become animal feed, Tesco's food waste equalled about 25% of that produced by all UK grocery retailers in 2014.

Suppliers and customers

It's likely that those 59,400 tonnes of food waste makes up only a small proportion of the food that is wasted in Tesco's supply chain or after it's been sold.

Supermarkets produce a relatively small proportion of the UK's food waste through their own operations. Households throw away just under half of the UK's food and drink waste, according to WRAP's estimates, whilst food manufacturers waste about a quarter. Grocery retailers and wholesalers, which includes supermarkets, account for less than a fiftieth of the total.

Tesco runs programmes designed to reduce food waste by its suppliers and customers. And it's signed the Courtauld Commitment to help reduce food waste across the UK food industry as a whole.

26 July 2016

⇨ The above information is reprinted with kind permission from Full Fact. Please visit www.fullfact.org for further information.

How the UK's household food waste problem is getting worse

As many as 8.4 million families in the UK struggle to put food on the table – and yet, 7.3 million tonnes of food waste ends up in landfills each year. Hazel Sheffield looks at how the food industry can help achieve the food waste goals the Government has missed.

The UK Government has failed to meet its target on household food waste. Statistics from the Waste and Resources Action Programme (WRAP) show household food waste in the UK increased 4.4 per cent between 2012 and 2015, despite a target to cut household waste 5 per cent by 2015. That brings the amount of food sent to landfill in 2015 to 7.3 million tonnes, costing UK families £700 per year.

The numbers are disappointing after huge government-funded efforts to get UK households to stop throwing food in the bin. In the last ten years, campaigns like 'Love Food, Hate Waste' have been designed to make people rethink the way they shop and cook. Charities have heaped pressure on supermarkets to improve packaging so that consumers know when food is no longer edible. The Government even established a voluntary contract called the Courtauld Agreement between supermarkets and manufacturers to stop food from going to landfill. It's precisely this target that was missed.

The rise in food waste is shameful at a time when 8.4 million UK families are struggling to put food on the table, the equivalent of the whole population of London. As many as half of these families regularly go a whole day without food. The UK is in the bottom half of European countries for food insecurity, even as we keep throwing more and more food away.

Who's to blame?

Marcus Gover, CEO of WRAP, said the statistics show every person in the UK is responsible for reducing food waste. "Every person in the UK can help reduce food waste. WRAP's research found that almost 60 per cent of people believe they personally waste either no food or hardly any, even though people know food waste is an issue," he said.

However, food waste activists The Real Junk Food Project said the food industry, not the consumer, must take responsibility.

"I'm sick of the food industry passing their responsibility onto the consumers as though it's our fault," said Adam Smith, founder. "More and more people are composting than ever before. More and more people have less food than ever before and this Christmas period, food sales were down due to a 14 per cent increase in food prices for Christmas lunch."

Statistics

WRAP said numbers were partly to blame for the missed target. It described the increase in household food waste as "not statistically significant", meaning it doesn't think household waste has really increased if you take into consideration population growth, falling food prices and greater earnings.

Dominic Hogg, chairman of Eunomia, an environmental consultancy, agreed that it was hard to draw conclusions from the data. "Estimating the amount of household food waste in the UK is not easy to do in a way that reliably tracks small movements in the quantities of food waste," he said, noting that any apparent increase in food waste might have been higher without the Courtauld Agreement.

WRAP said that real earnings fell between 2006 and 2014, meaning families may have tried to rein in food waste to cut the cost of the supermarket shop. But in 2015, earnings staged a recovery and rose 1.4 per cent.

During this time food prices were increasing faster than other goods, with food price inflation averaging 5.1 per cent between 2007 and 2013. Then at the end of 2013 the numbers reversed, with food price inflation decreasing and eventually turning negative, meaning food started to get cheaper each year. Faced with more money and lower prices, households may have had less of an incentive not to waste food, WRAP said, though it could not say for sure just how much this affected the figures.

Food waste collections

One part of the UK did successfully reduce household food waste: Wales. In Wales, there is evidence that household food waste levels reduced 12 per cent between 2009 and 2015 and are now lower than the rest of the UK by around nine per cent.

Dr David Moon, head of food sustainability at WRAP, said there are two reasons for Wales's success. First, average earnings in Wales are lower than the UK average. "National statistics show average weekly earnings are ten per cent lower in Wales than in the rest of the UK," Dr Moon says. "People tell us that saving money is one of the strongest motivations with putting food in the bin. So we see this association between less waste and saving money."

The Welsh Government has also adopted more ambitious targets on food waste in the UK. Some 90 per cent of local authorities collect food waste in a separate caddy in Wales, compared to just 27 per cent in the UK. "If I am in the kitchen and I'm putting all my waste into a caddy, I'm more aware of how much I'm throwing away and more likely to do something about it," Dr Moon says.

Taking responsibility

The Welsh case suggests the UK Government could have a huge

impact on household food waste. A joined-up policy across local councils would ensure that every household had access to separate food waste collections.

While people make their own decisions about what to throw away, supermarkets are in contact with millions of customers every week, giving them the opportunity to work as a force for good in reducing food waste, rather than encouraging customers to buy more than they need with wasteful offers.

Some supermarkets are better than others. Sainsbury's has invested millions in reducing household food waste. Last year, Sainsbury's removed multibuys in favour of simpler pricing so that customers are no longer encouraged to buy more than they need. The supermarket has also dedicated £10 million to a five-year programme designed to help customers save money by throwing away less food, including a £1 million investment to turn the Derbyshire town of Swadlincote into a testing ground for new ideas to cut waste.

"We know we can't fix the problem single-handedly, however, which is why we are calling on other retailers, non-governmental organisations and the food industry so we can work together to help solve the problem," a spokeswoman said.

The UK now has until 2025 to meet the next Courtauld Agreement target of reducing food and drink waste by an ambitious 20 per cent. That will only be achieved if supermarkets, manufacturers and local authorities are unified in their efforts put food sustainability before profit. Food waste might come from the home, but tackling it calls for a sea-change in the way the country thinks about food that starts way before shoppers leave the supermarket.

January 2017

⇨ The above information is reprinted with kind permission from *The Independent*. Please visit www.independent.co.uk for further information.

Getting to grips with our food waste

The UK's food waste problem is galvanising agreement between retailers, policymakers, NGOs, activists, campaigners and consultants, says research from the Sustainable Consumption Institute (SCI) at the University of Manchester.

One third of the food currently produced for humans to eat is wasted. Until recently, responsibilities for food waste were pushed onto the shoulders of individuals and households. "Initial responses either blamed consumers for their assumed lack of concern and cooking skills or else positioned food waste as a matter of consumer choice and behaviour change," says project leader Dr David Evans.

But research by SCI shows that the issue is much more complex. A survey of 2,800 consumers shows that conventions generally seen as positive – cooking from fresh ingredients, the nuclear family meal, and eating with friends and extended family – are more likely to give rise to surplus food, which is at risk of becoming waste.

Moreover, since 2012, austerity, food price inflation, food poverty, the growing use of food banks, and concerns about food security, environmental sustainability and climate change, have all come together to transform responses to the challenge of food waste reduction.

"Our research suggests that a surprising multi-stakeholder coalition has rapidly developed around this issue and there is broad consensus among retailers, activists and policymakers," explains Dr Evans. "Crucially, there is recognition that we must move beyond 'behaviour change' and recognise, and advocate for, distributed responsibility."

Supermarkets, who are directly responsible for only five per cent of food waste, are taking measures to help suppliers and customers waste less. Current initiatives include the introduction of guaranteed minimum orders for suppliers, better redistribution of surplus food to charities, improved packaging and a clearer focus on helping consumers to avoid waste.

The reasons for supermarkets getting on board are complex, but they do believe customers care about food waste and inactivity could lead them to shop elsewhere. Among all stakeholders, researchers identified a sense of moral outrage regarding food waste and a genuine shared desire to tackle the problem.

"It will be interesting to see if this move beyond behaviour change and the cohesion between parties that are usually rather antagonistic is indicative of broader change in how retailers and their stakeholders approach other social or environmental issues," says Dr Evans.

20 July 2016

⇨ The above information is reprinted with kind permission from the ESRC. Please visit www.esrc.ac.uk for further information.

Minutes on the lips, a lifetime on the tip: the coffee cup waste mountain THE CONVERSATION

***An article from* The Conversation.**

By Jennifer Ferreira, Senior Research Assistant, Coventry University

In the UK we have got used to getting "what we want, when we want it". So began Hugh Fearnley-Whittingstall's latest instalment of his *War on Waste*. And he continued: "the cost of consuming so much stuff is waste… mountains of it". Last year witnessed his campaign for supermarkets to reduce food waste, and this year he has turned his attention to lack of recycling of coffee cups and the excessive packaging used by some online retailers.

It won't have escaped anyone's notice that the number of coffee shops in our towns and cities has grown rapidly over the past decade. As a consequence the number of disposable cups thrown away has exploded to an estimated 2.5 billion a year, or around 5,000 every minute.

The sheer number of cups is clearly staggering, but the real controversy covered by Fearnley-Whittingstall's programme was that these cups are not being recycled – even though the companies claim to be making efforts to do it.

Drunk and dunk

Many consumers assume that coffee cups can be recycled. In fact, it is not that simple. Most coffee cups are lined with a layer of polyethylene which makes the cup waterproof, but makes it difficult for the cup to be effectively recycled, because it can't be easily separated. There are few recycling facilities that can process this kind of cup.

Even though some coffee chains say that they have largely recyclable cups, the evidence presented was that most cups were not sent for recycling. There are estimates that fewer than one in 400 cups get sent for recycling. The TV show suggested that only one facility in the UK had the capability to recycle some of the chain coffee shop cups, but that very few were actually even sent there. Overall it is estimated that 0.1% of cups are recycled.

The upshot is that millions of consumers each week are putting their coffee cups into recycling bins and they are not being recycled. It is misleading for the thousands of people every day who thought they were doing their bit. The consequence of this, according to Fearnley-Whittingstall, is that the public think they are already recycling and therefore don't know there's a problem – so they don't ask for a solution.

High Street dominance

As with many celebrity-backed TV campaigns, this one has received significant attention both on social media, and in the newspapers, with the large high street coffee shop chains being lambasted for their recycling failures.

The coffee shop industry is one of the most successful retail sectors in the UK. There are more than 18,000 outlets today – and industry estimates suggest this could grow to more than 27,000 by 2020. That signals a troubling trend for more and more disposable coffee cups to end up in our landfill sites. Unsurprisingly, the main targets of the War on Waste coffee cup campaign were the big coffee shop chains: Costa, Starbucks and Caffè Nero.

There have been some responses, in particular from Costa, which publicly responded on Twitter to say the company acknowledged the issue and welcomed innovation in the area of takeaway cups. It also published infographics to indicate its commitment to addressing the problem.

The issue of disposable cups as an environmental problem is not new and has been discussed for many years within the industry. And it is still possible that steps can be made to make better recyclable cups and improve the process by which they can be recycled. But there are other efforts that can be made too.

Estimates from retail and hospitality consultant Allegra Strategies suggest that one in five people visit a coffee shop on a daily basis (up from one in nine in 2009).

That means there is a huge population of coffee drinkers which, if it changed its consumption habits to carry a reusable coffee cup, could make a massive impact on disposable cup usage. There is an increasing range of reusable coffee cup options on the market. There are those offered in the chains themselves, but also from other providers – Keep Cup and Joco Cup are just two examples.

Businesses can encourage this process too, as some have been doing; for example, Boston Tea Party, which offers discounts on drinks when people use reusable cups, and through other efforts to work towards more sustainable solutions, such as recycling coffee grounds or sustainable building projects. Business can also make coffee cup recycling more mainstream in the UK, building on pledges made as part of the Paper Cup Manifesto backed by a range of coffee shop industry stakeholders.

Fearnley-Whittingstall stated on the show that "waste is an environmental disaster we can all do something about" – and, if the coffee shop industry can make efforts to provide more recyclable cups, the public should think about its role too. We should not only be pressuring these companies on Twitter, but addressing our own habits in a concerted effort to wage this "War on Waste".

29 July 2016

⇨ The above information is reprinted with kind permission from *The Conversation*. Please visit www.theconversation.com for further information.

A beginner's guide to the waste hierarchy

Much of how we handle waste in the modern world is now driven by what is called the waste hierarchy. It's a process that is outlined in the EU Waste Framework directive and involves the important stages of prevention, reuse and recycling, before final disposal. The aim of the hierarchy is to make sure that we send much less of our waste to landfill and recycle as much as possible.

In the UK alone we produce in the region of 100 million tonnes of waste annually. Every home and every office and industry produces its fair share of rubbish. Each household creates about a tonne of waste every year and the average home throws away the equivalent of six trees worth of paper and card. The good news is that almost two thirds of the paper waste that we produce in the UK is now recycled thanks to local councils and their waste management processes.

Sticking to the waste hierarchy is a key factor in all this.

Waste prevention

The best way to cut down on waste is to prevent it in the first place. This is the immediate focus of the waste hierarchy and applies as much to households as it does to offices and corporations. While home owners have less legislation to contend with when it comes to handling waste, we all have a duty of care to make sure that we don't produce it in large amounts. Making a few lifestyle changes can greatly reduce the amount of waste that your household produces each year.

In the home this could include:

⇨ Using up left-overs for meals rather than throwing into the waste.

⇨ Not buying more than we need, particularly when it comes to doing the weekly shop.

⇨ Opting for products that have less packaging or biodegradable packaging.

⇨ Reducing the amount of junk mail you receive by moving to digital notifications or asking for mail deliveries to be stopped.

In the office, there is even more scope for preventing waste:

⇨ Reducing the amount that is printed out by using smart devices to view files and information.

⇨ Choosing suppliers who provide recycled products and stationery.

⇨ Introducing an office policy to help employees reduce waste.

⇨ Using less packaging with products and reusing material such as envelopes.

Preparation for reuse

The next option after preventing general and other waste is to reuse what you have already. This can be something as simple as using two sides of paper before sending it to waste or refurbishing equipment that can either be sold on or reused. If you're having a clear out at home, this would entail sorting out material that can either go to the local charity shop or to friends or can be sent for recycling. In the office, it can mean that when hardware needs to be replaced it is cleaned up and sold on rather than being sent straight to landfill.

Recycling of waste

The next step in the waste hierarchy is recycling. We have become a lot better at this over the last decade or so as laws and processes have come into place to help. This includes most of us having recycling bins that can be collected by local councils and large facilities that take material such as paper, plastic and metal and transforms them into new products. Recycled paper is now much more prevalent than it used to be and is produced at a lower cost than paper made from 'virgin' sources.

Other processes

When material cannot be reused or recycled, then there are a few other options to landfill that are available nowadays. This includes the incineration of waste to produce heat and power that can be used for the local community. There are also processes such anaerobic digestion which are becoming popular, particularly for food and agricultural waste.

Disposal

The key to the waste hierarchy is ensuring that we send less and less to landfill. For all materials that cannot be disposed of any other way this is the final solution. In the UK we have dramatically reduced the amount of waste we collect and send to landfill sites. Specialist licences are need to operate this sort of facility and the type of waste that ends up here is highly regulated.

Follow the waste hierarchy and you can help make a real impact on the environment. Businesses have to do this by law but we should all, as householders, keep the hierarchy in mind when we come to sort out our waste.

13 June 2016

⇨ The above information is reprinted with kind permission from the Waste Disposal Hub. Please visit www.wastedisposalhub.com for further information.

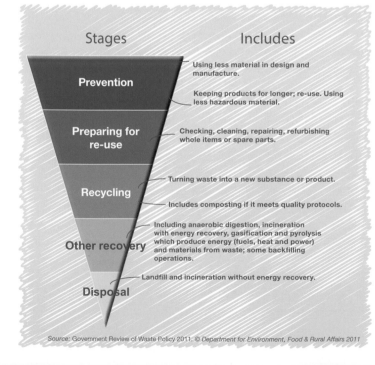

Stages	Includes
Prevention	Using less material in design and manufacture. Keeping products for longer; re-use. Using less hazardous material.
Preparing for re-use	Checking, cleaning, repairing, refurbishing whole items or spare parts.
Recycling	Turning waste into a new substance or product. Includes composting if it meets quality protocols.
Other recovery	Including anaerobic digestion, incineration with energy recovery, gasification and pyrolysis which produce energy (fuels, heat and power) and materials from waste; some backfilling operations.
Disposal	Landfill and incineration without energy recovery.

Source: Government Review of Waste Policy 2011. © Department for Environment, Food & Rural Affairs 2011

Can incineration and landfills save us from the recycling crisis?

In many parts of the US, it now costs more to recycle old plastic than to make new products. So what are the alternatives for getting rid of our waste?

By Bruce Watson

It's been a tough year for plastic recycling, and the culprit is oil.

Over the past two years, petroleum prices have plummeted, at one point dropping to 70% below June 2014 levels. As prices have fallen, they've dragged down the cost of virgin plastic, which is made from oil. In many areas, it now costs more to recycle old plastic than to make new containers.

Environmentally, there's no question that recycling is the best method for dealing with waste. Recycling one ton of aluminium saves 14,000kWh of electricity – compared to making aluminium from raw materials – more energy than the average household uses in a year. Paper products are less profitable, but recycling one ton of cardboard still saves 390kWh – more than a week's worth of electricity.

Some commodities are highly profitable: scrap aluminium, for example, is worth $1,491 per ton. Paper or cardboard sells for $90–$140 per ton. This nets recyclers a handy profit, even after processing costs.

When plastic was more expensive, recycling helped offset the expense of recycling less-profitable materials, like glass. But as the value of plastic has dropped, it has had ripple effects across the recycling industry. Waste Management's recycling division posted a $16 million loss in the first quarter of 2016, and the company has shut almost 30% of its recycling facilities. Meanwhile, questions about the treatment of recycling workers and the large amounts of recycled glass

and plastic that still go to landfill have tarnished its reputation.

The other problem is that many places only collect one or two types of plastic, instead of all products that could be processed. And areas that collect all plastics sometimes end up sending many types to landfills, even after consumers recycle them, because the returns are too low to make recycling it all economical.

But, as recycling costs have gone up, a combination of technological advancements and increased environmental regulation have made other disposal options increasingly viable. Here's how three common waste management options stack up.

Incineration: waste to energy

Waste-to-energy (WTE) trash incineration, which burns waste to generate electricity, is a promising option. In addition to disposing of garbage and reducing landfill space, WTE generates 500kWh of electricity per ton of waste – roughly the same amount of power generated by a third of a ton of coal.

For all these benefits, however, WTE plants are rare: there are only 84 WTE facilities currently operating in the US. Florida's Renewable Energy Facility Two, which opened last year, was the first US WTE plant to open in the past 15 years.

Part of the problem is that WTE plants are costly to construct, and companies often offset this by negotiating long-term contracts with cities. "Cities get

locked into a contract and can end up on the hook for huge fees to waste processors, regardless of whether or not there is enough waste for them to process," says Monica Wilson, US and Canada programme director at Gaia, a worldwide alliance of non-profit organisations that fights waste-to-energy garbage incineration.

Pollution is another concern. "Whether dioxin, mercury, lead and other toxins go out the stack, are captured, or end up in the ash that is left over after incineration – they're still there," Wilson says.

Nickolas John Themelis, an engineering professor at Columbia University and chairman of the Global Waste to Energy Research and Technology Council, argues that pollution concerns are overblown. "Studies have shown that the entire US WTE industry produces three grams of dioxin per year," he says. "By comparison, there are over 3,000 landfill fires reported every year, and they produce 1,400 grams of dioxin."

As for the high costs of WTE plants, Themelis argues that a large part of the expense is caused by critics, whose protests and lawsuits can – he says – tip the scales and make the technology unprofitable. "These plants are very expensive to build, and years of litigation by a very vocal minority can make it too expensive."

Even so, many communities that are committed to reducing waste continue to use WTE. Instead of building new plants, some quietly ship their waste to existing facilities: in 2014, for example, New York City committed to send 800,000 tons of trash to a facility in New Jersey.

Burying the problem: landfill

Landfills are the most common and economical waste management solution. Their cost varies widely across the country, averaging out to $48.27 per ton. This covers the entire lifetime cost of a landfill, from the purchase and preparation of the land to maintenance and monitoring.

Environmentally, of course, landfills have a terrible reputation, and have been cited for problems including groundwater contamination and air pollution. While the Environmental Protection Agency (EPA) passed rules regulating their environmental impact in 1991 and 1993, a large part of the problem is consumer behaviour. "Many of the things we see in landfills could have been reused or recycled, but consumers didn't put them into the reuse or recycling waste streams," says Anne Germain, the director of waste and recycling at the National Waste and Recycling Association.

Landfills can also produce energy. In 1996, the EPA passed laws requiring large landfills to capture their gas emissions. "Landfill gas" – which contains methane, CO_2 and about 30 hazardous organic compounds – can be used as an alternative to fossil fuels to produce heat and electricity. According to the EPA, 648 of the 2,400 municipal solid waste landfills in the US have one or more landfill gas collection projects attached to them. Some 400 more could cost effectively be used for generating methane; if they were, the EPA says, they could power 473,000 homes.

Germain argues that landfills represent a realistic, promising solution to waste processing.

"The idea that we can divert all of our waste is a dream," she says. "It's not going to happen soon. In the meantime, landfills have to exist to catch the things that we can't take out of the waste stream."

Digestion: swallowing our waste

For communities that produce a lot of organic waste, anaerobic digestion offers an environmentally sound solution. Tim Flanagan, general manager at a Monterey, California anaerobic digestion plant, compares the process to a giant crock pot: "We put in a mix of material – about 75% food waste and 25% organic yard waste – and let it cook for 21 days." He continues: "It produces methane, which we use to run an engine generator. We have a net yield of about 80kW, which helps power a nearby sewer agency."

After fermentation, the leftovers go to an on-site composter, where they are turned into a fertiliser that is sold to local farms and vineyards.

The programme charges $51.75 per ton, with discounts for separated organic material and food wastes. Flanagan says the digester costs slightly more than landfilling, but it also uses far less space and produces less pollution and more energy. The digestion also takes place in a controlled environment, where all the gas it produces can be collected.

The facility's next step is to begin compressing its methane, which could then fuel its garbage trucks. "We'll have trucks picking up food waste, and digesters turning it into fuel to run the trucks, so they can pick up more waste," Flanagan says.

14 June 2016

⇨ The above information is reprinted with kind permission from *The Guardian*. Please visit www.theguardian.com for further information.

Waste legislation and regulations

Guidance for businesses and organisations on how waste disposal is regulated and what they need to do to comply.

Relevant legislation and regulations

The EU Waste Framework Directive provides the legislative framework for the collection, transport, recovery and disposal of waste, and includes a common definition of waste . The directive requires all member states to take the necessary measures to ensure waste is recovered or disposed of without endangering human health or causing harm to the environment and includes permitting, registration and inspection requirements.

The directive also requires member states to take appropriate measures to encourage firstly, the prevention or reduction of waste production and its harmfulness, and secondly, the recovery of waste by means of recycling, re-use or reclamation or any other process with a view to extracting secondary raw materials, or the use of waste as a source of energy. The directive's requirements are supplemented by other directives for specific waste streams.

The Waste (England and Wales) (Amendment) Regulations 2012 were laid before Parliament and the Welsh Assembly on 19 July 2012 and came into force on 1 October 2012. The amended regulations relate to the separate collection of waste. They amend the Waste (England and Wales) Regulations 2011 by replacing regulation 13. From 1 January 2015, waste collection authorities must collect waste paper, metal, plastic and glass separately. It also imposes a duty on waste collection authorities, from that date, when making arrangements for the collection of such waste, to ensure that those arrangements are by way of separate collection.

These duties apply where separate collection is necessary to ensure that waste undergoes recovery operations in accordance with the directive and to facilitate or improve recovery, and where it is technically, environmentally and economically practicable. The duties apply to waste classified as waste from households and waste that is classified as commercial or industrial waste. The amended regulations also replaced regulation 14(2) to reflect the changes to regulation 13 to ensure a consistent approach. Consequential changes are also made to reflect changes in paragraph numbering in the new regulation 13.

Our combined summary of consultation responses and government response to the consultation on amending the Waste Regulations 2011 on the separate collection of recycling has also been published.

Environmental permitting for waste

The recovery and disposal of waste requires a permit under EU legislation with the principal objective of preventing harm to human health and the environment. This legislation also allows member states to provide for exemptions from the need for a permit, providing general rules are laid down for each type of exempt activity, and the operation is registered with the relevant registration authority. We have given effect to the EU requirements through the Environmental Permitting (England and Wales) Regulations 2010 (the 2010 regulations). More information is available on the National Archive and on the Environment Agency website.

Hazardous waste regulations

Hazardous waste is essentially waste that contains hazardous properties which, if mismanaged, has the potential to cause greater harm to the environment and human health than non-hazardous. As a result, strict controls apply from the point of its production, to its movement, management, and recovery or disposal.

Waste shipment regulations

Waste shipment regulations are comprised of EU Regulations, a UK statutory instrument and a UK Plan. Between them, they control movements of waste between the UK and other countries and provide a framework for enforcement. Some movements are prohibited, others are subject to prior written notification and consent procedures and some are subject to basic administrative controls. The control depends on the nature of the waste, its destination and whether it is destined for recovery or disposal. You can find more information on the National Archive.

UK ship recycling strategy

Ship recycling is a global issue. Defra considers the environmentally sound management of ships to be a high priority and in 2007 issued a Ship Recycling Strategy for UK ships.

Electrical and electronic equipment

Waste Electrical and Electronic Equipment (WEEE) and Restriction of Hazardous Substances in electrical and electronic equipment (RoHS) directives aim to reduce the quantity of waste from electrical and electronic equipment and increase its re-use, recovery and recycling. The RoHS directive aims to limit the environmental impact of electrical and electronic equipment when it reached the end of its life. It does this by minimising the hazardous substances of legislation controlling hazardous substances in electrical equipment across the community. More information is available on the Environment Agency website.

Packaging, packaging waste and packaging waste regulations

These regulations aim to harmonise national measures concerning the management of packaging and packaging waste to provide a high level of environmental protection and to ensure the functioning of the internal market. For more details read the Government's policy on reducing and managing waste.

Landfill directive

This directive aims to prevent or reduce as far as possible negative effects on the environment from the landfilling of waste, by introducing stringent technical requirements for waste and landfills and setting targets for the reduction of biodegradable municipal waste going to landfill. For more information, read the Government's policy on reducing and managing waste.

End-of-life vehicles (ELVs) Regulation 2003

This regulation aims to prevent waste from end-of-life vehicles and promote the collection, re-use and recycling of their components to protect the environment. More information is available on the Environment Agency website.

Batteries directive

This directive aims to improve the environmental performance of batteries and minimise the impact waste batteries have on the environment. It does this by:

⇨ restricting the use of cadmium and mercury in the design and manufacture of new batteries

⇨ setting collection and recycling targets for waste portable batteries

⇨ banning the disposal of untreated automotive or industrial batteries in landfill or by incineration.

Last updated 2014

⇨ The above information is reprinted with kind permission from GOV.UK.

Clean and inclusive? Recycling e-waste in China and India

Extract from the report by IIED.

By Kate Lines and Ben Garside, with Satish Sinha and Irina Fedorenko

E-waste is one of the world's fastest growing waste streams. We generate over 40 million tonnes annually. By managing e-waste well, valuable secondary raw materials can be recovered. There are also significant greenhouse gas emissions savings from associated displaced mining activities. Refurbishing reusable parts makes possible cheap access to household appliances and communications technology, which is particularly important for low-income rural consumers in developing countries. Better e-waste management can play an important role in the transition to greener and more inclusive economies. But much of e-waste's potential as a resource is lost when it is improperly processed. Electrical and electronic equipment (EEE) contains a range of toxic materials that can be harmful to human health and the environment, and intensify the vulnerability of workers.

Why explore e-waste management in India and China?

This paper explores issues of e-waste management in China and India: two significant players in e-waste generation and recycling. China is second only to the USA in total domestic generation (4.4kg per capita in 2014 compared to 22kg in USA). India stands fifth globally, generating 1.6 million tonnes (a tiny 1.3kg per capita). Over the last few decades, illegal international import 'dumping', poverty and urbanisation combined with non-existent or weak regulations have created the conditions for the growth of huge, informal and unregulated e-waste sectors in these countries. These operate at all stages of the chain – from collection, dismantling and trade to repair, recycling and metals recovery. Now, since 2012, for the first time growing domestic consumption has overtaken international imports as the main source for India and China's rising mountains of e-waste. New governance challenges are emerging in a range of specific domestic contexts, demanding national and local policy attention beyond just integrating or complying with international toxic waste import/export regulations.

What is the problem?

Driven by laudable environmental and health concerns, both countries are stepping up efforts to regulate.

But attempts so far to encourage the development of a formal recycling sector and 'clean channels' are having little impact on the informal sphere or on national e-waste markets. Both countries feature thriving informal e-waste markets that are inclusive of some of the poorest, and which deal with the overwhelming majority of imported and domestic e-waste – in India this could be as high as 95 per cent.

While these informal operations are not often green and can be exploitative, they represent many thousands of livelihoods, and informal practices often have significant efficiency gains over formal counterparts, particularly in the early stages of the chain (collecting, dismantling and refurbishing). Formal e-waste management supply chains have proved more expensive to establish and operate – in part because they must observe exacting environmental and health standards. Regulations in India have banked on the 'extended producer responsibility' (EPR) principle to push the burden on to EEE producers, but so far there has been little widespread implementation on the part of the producers, or enforcement by state regulators. And while the Chinese Government is seeking new and innovative ways to formalise, centralise and clean up systems for e-waste collection and recycling, most recent attempts have not got far in transforming the market.

What are the challenges and opportunities for policymakers?

There is a clear mismatch between regulation and reality in both countries, where policy preference has drawn on European and US models (themselves often flawed), but has taken them out of context. Indian and Chinese policymakers' preference is to create conditions of formal sector competition with well-established and dominant informal markets, largely ignoring or trying to further criminalise and marginalise informal e-waste enterprises – putting at risk many informal-sector livelihoods.

Policymakers should instead see the opportunity for solutions that build on established supply chains and collection practices to harness the benefits and efficiencies of the informal networks already in place. In this paper, we present a case for integrating and building on existing realities. These realities include not only a large informal sector that has matured in a relative regulatory vacuum, but also regulatory bodies' weak capacity for enforcement (relative to their European counterparts), and household consumption habits grounded in long-standing cultural practices. Market governance that ignores these realities will present few prospects for environmental improvements and lead to ineffective incentive structures, either for new formal businesses to establish or for existing informal enterprises to clean up their operations, perhaps through formalising.

What models might work?

What interventions could better green the e-waste sector while being inclusive of existing informal small and medium enterprises? In this paper we reflect on current policy and practice across a spectrum of examples that encompass formal and informal sector players. We discuss the challenges and opportunities for improved environmental and health outcomes across these two countries' e-waste sectors – and the degree of inclusion different potential e-waste models provide. One solution may be hybrid models, where existing informal collection and dismantling markets (in particular) are supported by, and integrated with, formal recycling companies that use clean technologies for metal recycling and toxic compound disposal. We tentatively explore the potential of a few specific models, focusing on India.

At a conceptual level, there is a strong business case for hybrid formal–informal models. We know that the informal e-waste sectors in China and India are huge.

Ignoring them or making them more illegal with punitive regulation is not likely to be effective in getting e-waste flowing into clean recycling channels. The process of inclusive formalisation of wastepickers more broadly is also certainly not new, in India and other places – recent examples include Solid Waste Collection and Handling (SWaCH) in Pune, India; National Waste Pickers Movement in Brazil; Bogota Waste Pickers Association in Colombia; and the zabaleen in Cairo, Egypt.

What are the main lessons?

Based on the analysis, we can draw seven broad conclusions. These point to critical challenges in governing the e-waste sector inclusively and to mechanisms which could steer e-waste markets towards greener and fairer outcomes. Exploring these further will help build an understanding of how hybrid models could work in practice, including the types of supporting policies that would help with adoption and scale. While they are directed particularly at China and India – two countries processing a very large proportion of the world's e-waste – they also suggest useful lessons for other countries or cities where policymakers are grappling with similar challenges of how to manage the growing mountains of domestically-generated e-waste already claimed by thriving informal markets.

1. Regulation must be inclusive, building on existing realities: e-waste governance must consider those individuals and enterprises, particularly the poorest or smallest, already working in the sector – and their existing practices. Appropriate regulatory incentives should enable rather than disable them to engage in greener practices. Policymakers should recognise the efficacy and diversity of the informal economy, but also existing structural and power inequalities.

2. The economics, institutions and incentives must work: effective interventions to address livelihood, health and environment issues together must also address the realities of pricing for e-waste within current established informal markets. This is also true for the success of formal sector channels, whether or not they aim to encourage an inclusive sector. Economic and other incentives, including buying in to different organisational model options along the e-waste value chain, must be thought through and discussed with actors along the whole value chain.

3. Refurbishment is greener, and a critical part of current informal sector models: policy and regulation must recognise the importance of reusable e-waste, for the informal economy and for low-income consumers, and for the environment.

4. Municipalities can play a key role in working with the informal sector, tackling spatial exclusion by providing valuable and much needed space for income-generating activities and helping to lower barriers to formalising. Local authorities are key stakeholders and stand to benefit significantly from efficient solid-waste – including e-waste – management.

5. The public needs better information: better information and education channels are vital to influence public behaviour. Existing consumer attitudes, household/office practices and cultural norms also need to be factored into policy design.

6. Engage electronic equipment manufacturers and retailers to promote experimentation and innovation of more inclusive e-waste management models. This must move beyond rhetoric to practical participation and experimentation in multistakeholder models – with ambitions of going to scale.

7. Developed country governments still have a responsibility and must pay increased attention to international law enforcement to prevent illegal shipments of e-waste to the developing world.

Fostering and scaling-up hybrid models is not an easy task. Models of formal–informal integration aiming for environmental protection or economic efficiency will not automatically lead to social welfare synergies. We need to consider carefully what incentives diverse stakeholders need to participate – and what are the potential barriers and inequalities. Hybrid models must recognise and avoid many potential risks. How can they be designed to prevent intensifying further vulnerabilities? Models will need to manage power and voice imbalances, prevent exploitation due to diverging interests, and ensure that informal worker organisations are not undermined. In addition, two of the biggest – and linked – challenges for designing greener and more inclusive models remain critical: appropriate pricing incentives for e-waste to flow into safe and clean recycling channels and the loss of livelihoods for those involved in hazardous recycling practices.

While the nature of e-waste production and disposal is of global concern, the issues are becoming increasingly complex. They go far beyond environmental injustice in developing countries. Dealing with the polluting nature of e-waste management techniques is urgent – as is the need to recover materials. But moves to cleaner, greener futures must be equitable, protecting the most vulnerable and seeking synergies with measures towards poverty reduction and social protection.

March 2016

⇨ The above information is reprinted with kind permission from IIED. Please visit pubs.iied.org for further information.

⇨ The original publication can be found at: http://pubs.iied.org/16611IIED/

How can we create a world where plastic never becomes waste?

Today nearly everyone, everywhere, every day, comes into contact with plastics. Plastics have become the ubiquitous workhorse material of the modern economy – combining unrivalled functional properties with low cost. And yet, while delivering many benefits, the current plastics economy has drawbacks that are becoming more apparent by the day.

Significant economic value is lost after each use, along with wide-ranging negative impacts to natural systems. How can we turn the challenges of our current plastics economy into a global opportunity for innovation and value capture, resulting in stronger economies and better environmental outcomes?

As part of Project MainStream, the World Economic Forum and Ellen MacArthur Foundation, with McKinsey & Company as a knowledge partner, have come together under the guidance of a Steering Board with nine global CEOs, to answer this question. Our latest report *The New Plastics Economy: Rethinking the future of plastics* lays out a blueprint for an economy where plastic never becomes waste.

What's the state of plastics today?

Our current plastics economy can be seen as the archetypal 'linear' (take-make-dispose) system:

⇨ Significant value is lost: most plastic packaging is used only once; 95% of the value of plastic packaging material, worth $80–120 billion annually, is lost to the economy.

⇨ Substantial negative economic costs: the cost of negative externalities from plastic packaging (e.g. on oceans and

urban infrastructure), combined with the cost associated with greenhouse gas emissions from its production is conservatively estimated at $40 billion annually – exceeding the plastic packaging industry's profit pool.

⇨ Under business-as-usual conditions, these negative costs are expected to increase significantly over the coming decades. Given the projected growth in consumption, by 2050 oceans are expected to contain more plastics than fish (by weight), and the entire plastics industry will consume 20% of total oil production and 15% of the annual carbon budget.

What does a New Plastics Economy look like?

At its core, it's a new way of thinking, underpinned by the principles of the circular economy. The New Plastics Economy is a vision of a global economy where plastics never become waste. It has three main ambitions:

1. Create an effective "after-use" plastics economy – improving the economics and uptake of recycling, scaling up the adoption of reusable packaging, and adopting compostable packaging for targeted applications, helping to capture lost value.

2. Drastically reduce the 'leakage' of plastics (e.g. plastics ending up in oceans) – by improving after-use infrastructure in targeted countries, increasing the economic attractiveness of keeping materials in the system, and reducing the negative impact of plastic packaging when it does escape the system.

3. Decouple plastics from fossil feedstocks – primarily through the development of renewably sourced plastics, i.e. either bio-based plastics or plastics made from captured greenhouse gas.

How do we get there?

Our report outlines a fundamental rethink for plastics; it offers a new, explicitly systemic and collaborative approach with the potential to transform global plastic packaging material flows and thereby usher in the New Plastics Economy. The reports major recommendations include:

1. Collaboration: establishing a Global Plastics Protocol and coordinating large-

scale pilots and demonstration projects, engaging all stakeholders across the value chain including policymakers and businesses.

2. Innovation: mobilising large-scale, targeted 'moon shot' innovations and continuing to develop and encourage scientific research in the area, such as the search for 'super-polymers' that combine functionality with superior recyclability, as well as bio-benign materials.

3. Communication: coordinating and driving communication on the nature of today's system, the vision of the New Plastics Economy, best practices, and insights.

While current innovations and efforts to improve the plastics economy show potential, to date these have proved to be too fragmented and uncoordinated to have impact at scale. These efforts need to be complemented and guided by a concerted, global collaboration initiative that matches the scale of the challenge and the opportunity. The Ellen MacArthur Foundation will establish such an initiative to act as a global dialogue mechanism and drive the shift towards the New Plastics Economy.

Those involved in the plastic packaging chain have proven time and again their capacity to innovate. Now, harnessing this capability to improve the circularity of plastic packaging – while continuing to expand its functionality and reduce its cost – could help us move towards a system that works: a New Plastics Economy.

The full report, *The New Plastics Economy: Rethinking the future of plastics* has been launched at the World Economic Forum's Annual Meeting in Davos.

Authors: Dame Ellen MacArthur, Founder, Ellen MacArthur Foundation, UK; Dominic Waughray, Head of Public-Private Partnership, Member of the Executive Committee, World Economic Forum.

19 January 2016

⇨ The above information is reprinted with kind permission from the World Economic Forum. Please visit www.weforum.org for further information.

Microbeads 'should be banned', MPs urge cosmetic companies to cease using product

"A single shower can result in 100,000 plastic particles entering the ocean."

By Kathryn Snowdon

Plastic microbeads, which are present in everyday household products such as exfoliating scrubs, toothpaste and shaving gel, should be banned, MPs have said.

Public awareness around the plastic product is growing, but many would still be unaware of the damage that can be caused – both to marine life and potentially human health, Parliament's Environmental Audit Committee said.

The cross-party committee is now demanding cosmetic companies are completely banned from using plastic microbeads.

"Trillions of tiny pieces of plastic are accumulating in the world's oceans, lakes and estuaries, harming marine life and entering the food chain. The microbeads in scrubs, shower gels and toothpastes are an avoidable part of this plastic pollution problem," Committee Chair Mary Creagh said.

"A single shower can result in 100,000 plastic particles entering the ocean. Cosmetic companies' voluntary approach to phasing out plastic microbeads simply won't wash.

"We need a full legal ban, preferably at an international level as pollution does not respect borders. If this isn't possible after our vote to leave the EU, then the Government should introduce a national ban.

"The best way to reduce this pollution is to prevent plastic being flushed into the sea in the first place."

Microplastic pollution comes from the fragmentation of larger pieces of plastic waste, small synthetic fibres from clothing, and the microbeads used in cosmetics and other products.

It is estimated that as much as 86 tonnes of microplastics is released into the environment every year in the UK from facial exfoliants alone.

The report, which was released on Wednesday, found that microplastic pollution was potentially more environmentally damaging than larger pieces of plastic because it is more likely to be eaten by wildlife.

Most large cosmetics companies have made voluntary commitments to phase out microbeads by 2020.

But the parliamentary committee urged a national ban to come into force by the end of 2017.

If a ban is not enforced, then products should be labelled clearly in order to provide transparency for customers, the committee urged.

Creagh said: "Most people would be aghast to learn that their beauty products are causing this ugly pollution.

"Cosmetic companies need to come clean and clearly label their products containing plastics."

Microbeads became popular in the 1990s and as a result of their small size, particles can travel through wastewater sewage treatments into the ocean, causing marine environmental damage, the MPs warned.

Professor Richard Thompson, from the University of Plymouth, cited a study that estimated that 680 tonnes of microbeads are used annually in cosmetic products in the UK.

The report highlights that Professor Thompson estimates that a single 150ml container of cosmetic product could contain about three million plastic particles – none of which are biodegradable.

There is also growing concern about the effect microplastic consumption can have on human health.

"Shockingly, a plate of six oysters can contain up to 50 particles of plastic.

"More research is needed on the impact of microplastic consumption on human health," said Creagh.

Although microbeads are a significant and avoidable part of the problem, the wider issue of microplastic pollution cannot be set aside once microbeads have been dealt with, the report found.

Between 80,000 and 219,000 tonnes of microplastics enter the marine environment across Europe per year.

Opportunities to capture microplastics through enhanced washing machine filtration systems and improved waste and water sewage treatment processes must also be explored.

24 August 2016

Fighting food waste with dumpster dinners

By Marie Mourad and Alex Barnard

By revealing and rescuing the piles of edible food left outside supermarkets, 'dumpster divers' have contributed to public mobilisations against food waste over the last few years. In the US, Barbara (all names are pseudonyms), a 45-year-old high-school Spanish teacher, sees dumpster diving as a way to focus media and popular attention on mass consumerism, which, in her eyes, is "destroying the planet." For Benjamin, a 25-year-old self-described anarchist and full-time activist, eating from the garbage is part of a strategy (alongside voluntary unemployment, squatting, bicycling and hitchhiking), for engaging in a "total boycott" of capitalism. Across the Atlantic, Pauline, a part-time French tutor and Quentin, who writes for a gastronomic guidebook, are a young Parisian couple organising free meals in their community garden with food rescued from the trash. They describe dumpster diving as "fun" and an "adventure."

From 2007 to 2016, our observations and interviews with dumpster divers and food waste activists in the United States and France, and our own embodied (and imbibed!) participation in recovering and re-using discarded food, have taken us beyond the spectacle of dumpster diving to its concrete practice. The series of practices for acquiring, preparing, and cooking with "waste" – as well as, occasionally, re-wasting it – reveals how the ethical commitment of turning waste into food creates challenges for adopting other ethical practices.

Acquisition: pavement shopping

In some surprising ways, dumpsterdiving – even when undertaken by anti-consumption activists – looks a lot like shopping. Regulars generally know where to go to find one type of food or another and they develop a weekly routine, built around the rhythm of stores putting out waste and municipal collection services coming to pick it up. Benjamin may not have bought food for over nine months, but he confidently rides his bike to what he knows as good "spots" and specifically looks for his favourite pizza or sandwiches. Like grocery-store shoppers who leave the house with a specific list but fill their carts with unintended purchases, he suddenly exclaims, "Whoa! There is sushi!"

Of course, divers can't be choosers, and 'shopping' in dumpsters does not quite have all the conveniences of consumption in modern capitalism. Divers regularly describe what they do as a modern day version of hunter-gathering, but it looks like fishing too. In opening a garbage bag, they cast a line into the vast waste stream of the metropolis, often with a half-joking, pseudo-religious plea to the "dumpster gods" – as Pauline calls them – to deliver a hoped-for item. Dumpster divers spend a lot of time waiting for garbage to appear and frequently boast about their findings, even the ones they had to let go.

By rejecting the importance of standardised, perfect-looking food and instead relying on senses of touch, taste, or smell, dumpster divers believe they have developed practices that are more sustainable than those of mainstream consumers in the current food system. They accurately pin partial responsibility for waste on supermarket sell-by dates and aesthetic standards (the former being potentially responsible for 30% of household food waste; the latter leading to the rejection of up to 20% of fruits and vegetables). Yet expectations about what constitutes so-called "good" food are not easy to change. Despite the growing interest in "ugly" products, even dumpster divers often opt for the least-blemished and most shapely items.

Preparation: the food-rescue recipe

Preparing a dumpstered meal can be almost effortless. For his part, Benjamin composes his diet from pizza, ready-made salads, sandwiches and, on special occasions, sushi. As Western consumers work longer hours and spend less time cooking, supermarkets have devoted an increasing amount of shelf space to pre-packed, pre-prepared foods. Although food packaging is often touted as a potential solution to consumer-level food waste, our catalogue of finds from over 60 dumpster diving expeditions tells a different story. We consistently found more cut pineapple than whole fruits, more individually packaged pasta salads than dry pasta; more cookies than flour, eggs or sugar. Some strategies manufacturers use to "add value" to their products cut into their shelf lives, leading inexorably to waste.

Most dumpster divers see food preparation as a way to break with more mainstream food practices, by reversing the usual assumption of the right way to cook: they know the individual ingredients that must be used, not what the final dish looks like. Cooking in this way requires creativity, improvisation, and knowledge of the material qualities of food. At times, though, the culinary traditions that dumpster divers think they are reviving, especially strict French gastronomic 'rules' about which vegetables must be peeled and which vegetables can be combined, do not mix easily with a commitment to reducing waste. In the end, taste and tradition have to be balanced against environmental concern.

Eating: garbage gourmands

Once the meal is ready, it is no longer 'waste'. Many dumpster-divers avow that they eat better food when dumpster diving than they ever would if they were buying it. Quentin and Pauline describe a diet full of choice cuts of meat, fancy cheese and foie gras – all delicacies that they could scarcely afford to actually buy. The abundance of dumpster-diving opportunities converts otherwise shoestring purchasers into garbage gourmands: "I've become very picky," Benjamin observes.

Still, the difficulty that even picky dumpster divers face in putting together healthy meals from the garbage raises questions about some of the strategies commonly proposed to reduce food waste. Reports often sum up the total calories that are discarded and then proceed to infer how many people could be fed by that food: by one calculation, current avoidable global food waste is enough to feed 1.9 billion people 2,100 kcal per capita per day. The implication is clear: that any excess food should be donated and used to feed the hungry. Yet a recent study from the U.S. Department of Agriculture found nearly half of the calories in discarded food come from added fats and added sweeteners. In recognition of this reality, Barbara once even cautioned new attendees at a trash tour about the risks of an all-dumpster-dived pastry diet, ruefully observing that she had gained weight since she started rescuing food.

The nature of the food procured by dumpster divers shapes the way food is eaten in often contradictory ways. Quentin and Pauline state that while they might prefer to be vegetarian, they feel a duty to eat the meat they find because of its high ecological footprint and economic value. But precisely because many divers see taking discarded food as having no environmental impact, they can consume these (former) commodities without reflecting on whether these meals and their preparation would make sense in a less wasteful world where they wouldn't dumpster dive. Dumpster divers have a much clearer view of how to live ethically as individuals within the existing food system than how to design a new system to replace it. Indeed, the temptation of effortlessly-recovered blueberry muffins or four cheese pizzas often draw them away from seemingly more constructive projects, like community gardens.

Disposal: "Don't waste the waste"

The very act of dumpster diving affirms the premise of social science 'discard studies': the act of disposing of something is not an end-point, but can actually begin a complex trajectory that – sometimes – ends in re-valuation and re-use. Having already saved their food and the natural resources and human labour embedded in it "from the jaws of the trash compactor," many divers are reticent to dispose of it again.

Yet the fact that the food is already rescued makes it first more likely, and second more legitimate, to let it to go to waste. One evening at Benjamin's apartment, Lucy, one of the temporary roommates and a regular dumpster-diver, opens the 'dumpster box', a cardboard box full of bread and pastries, and declares, "This bread is getting stale." Benjamin replies dismissively, "Just throw it out, we'll get more tonight." Lucy hesitates, and Benjamin adds, "You're just re-wasting it!" Unconvinced, Lucy defiantly takes a bite

with a loud 'crunch' and retorts "Don't waste the waste!" For Benjamin, like other groups of anti-waste activists, re-wasting is an unavoidable and necessary consequence of recovering food whose quality is open to debate. Quentin and Pauline's fridge overflows with items in various states of decay: they admit they may produce more waste than they did if they were not dumpster diving, partly because of the temptation of taking too much, and partly because of the greater proportion of packaged food.

At the core of some dumpster divers' critique is the notion that, in capitalist societies, useful things that cannot be sold are considered valueless. Yet dumpster divers themselves struggle to develop practices of care and stewardship for free things. As food banks and apps like Zero Percent and Phénix are promoted as a way to redistribute otherwise wasted food, these experiences caution us that not everything saved once from the dumpster escapes it forever.

Changing the world one dumpster at a time?

Irrespective of its diverse motives and goals, voluntary dumpster diving has had an influence – if not necessarily the one divers intended. One prominent UK food waste campaigner told us that, "By taking journalists round the back of supermarkets, showing them what was there, and serving them dinner based on it, and being able to very articulately talk about how this fits into a global problem," dumpster divers were among "the original instigators of this new wave of global action on food waste." But with reports from the European Union and United Nations declaring much of the 40% of world food production that goes to waste is discarded by consumers, attention has shifted to consumer practices – not "capitalism" or "industrial agriculture" – as a target for reform. As the United Nation's recently-launched Think.Eat. Save website assures us, "with relative ease and a few simple changes to our habits, we can significantly shift this paradigm" of waste.

Yet are changes to culinary practices as simple as activists and policymakers assume? We find that even dumpster divers – individuals often willing to renounce privileged

backgrounds and live on the margins of society, while facing the stigma attached to contact with waste – struggle to adopt truly sustainable practices. While their meals may be no impact in the sense that they have not financially contributed to the ecological or human costs of the food system, this is not the whole story. Some practices that go along with dumpster diving, such eating non-organic, non-seasonal, highly processed food and throwing away a large quantity of packaging, look surprisingly normal in their level of ecological consideration (or lack thereof). Ultimately, by focusing on individual practices, campaigns around food waste may be missing the extent to which these practices are constrained by the existing organisation of food production, distribution and consumption.

Further reading

Barnard, Alex V. 2016. *Freegans: Diving into the Wealth of Food Waste in America*. Minneapolis, MN: University of Minnesota Press.

Dubuisson-Quellier, Sophie. 2013. *Ethical Consumption*. Halifax, Canada: Fernwood Books.

Evans, David. 2014. *Food Waste: Home Consumption, Material Culture and Everyday Life*. London, UK: Bloomsbury Academic.

Mourad, Marie. 2016. "Recycling, Recovering and Preventing 'Food Waste': Competing Solutions for Food Systems Sustainability in the United States and France." *Journal of Cleaner Production* 126:461–77.

Stuart, Tristram. 2009. *Waste: Uncovering the Global Food Scandal*. New York: W. W. Norton & Company.

6 September 2016

⇨ The above information is reprinted with kind permission from The Discover Society. Please visit www. discoversociety.org for further information.

Britain's new wave of militant grocers

So much food goes into landfill while many families go hungry. Sarah Butler meets the shopkeepers fighting waste.

By Sarah Butler

A steady stream of shoppers are filling their baskets with biscuits, fizzy pop, risotto rice and tins of tomatoes at Niftie's grocery store in Dover on a cold January morning. Everything they're buying would have been thrown out by a mainstream supermarket. Shelves are piled with food well past its best-before date, slightly squashed cakes and some pretty unusual discontinued products including mojito mouthwash (90p), not usually available for sale in the UK.

If you're not bothered about instructions in another language, misprinted labels or biscuits that may be several months past their peak quality – but not stale – you can stock up for a fraction of the price you might pay in a regular shop. Prices start at 5p and go up to £1 with a jar of Ragu sauce selling for 50p and fruit juice for 40p.

Tracey Young, 53, a carer, single parent and Niftie's regular, says: "This saves me a fortune and they deliver when I buy too much to carry. I'm here every week and this shop has helped me out quite a lot. It's a shame so many supermarkets throw away good food."

Niftie's is part of a new wave of retailers trying to capitalise on a national avalanche of surplus food which is in demand from hard-pressed families.

The Real Junk Food Project, a charity which operates a chain of pay as you feel cafés using surplus food, now has two Sharehouse food stores connected to its operations in Leeds and Sheffield. Founder Adam Smith says he expects to open about ten more around the country by the end of this year. There are also several Affordable Foods stores in Cornwall as well as the national Company Shop group which redistributes surplus food through staff shops. The Sharehouse shops stock goods that are close to or past their best-before date as well as items that have been damaged or nabbed from supermarket bins. Some food is donated by retailers, including Sainsbury's and Marks & Spencer. Visitors make voluntary donations and some people have provided furniture for the café or volunteer their time in return for food.

Niftie's has just expanded online where it competes with Approved Food, a discount specialist that stocks 2,000 items which are surplus or close to their best-before date.

While the number of people using food banks has risen, there's plenty of surplus food to hand. Supermarkets and suppliers waste an estimated 1.9 million tonnes of food every year. In total, the avoidable food waste created by the grocery sector each year is worth around £1.9 billion.

Adam Smith says he believes even that is an underestimate. "It's shameful," he says. "Right now some people are in a desperate situation and there is outrage about food waste. We get a huge spectrum of the general public visiting. Some just want to support us and others need to have the food. There's a hidden level of poverty in the UK that can't claim benefits and can't afford to put good food on the table."

He reckons their Pudsey store gets between 500 and 1,000 visitors a day, with the café, school food programme and store all operating in partnership with each other to handle a mountain of surplus food.

Niftie's founder Nathan Richards, who began the store from his front room last summer, says he is operating as a non-profit social enterprise. A gifted market trader who began the shop as a way to get back into employment after struggling with depression, he is in a hectic whirl, chatting to customers, checking on a delivery of unwanted drinks and trying to pack up boxes for online orders. "The Government is letting the working class fend for themselves," he says. "We are stepping up and it shows British spirit. This business has had an effect on my town. More people are fundraising and are aware of others."

Richards gets his stock from supermarket suppliers who are stuck with too much produce after cancelled or reduced orders, or from the local food bank which doesn't like to give out food that's close to its best-before date. He goes for brands people can recognise, gets about a fifth of his stock for free and pays for the rest.

He admits it's tricky to make the economics stack up when most people spend only a few quid on a basket of items, but reckons he's moved 100 tonnes of stock since he opened in July, keeping it out of landfill and saving people money.

"My target is to help families who are struggling to get by. Nobody should be going without a meal."

26 February 2017

⇨ The above information is reprinted with kind permission from *The Guardian*. Please visit www.theguardian.com for further information.

Edible water: how eating little balls of H²O could be the answer to the world's plastic pollution

Plastic water bottles are one of the worst culprits for the 16 million plastic bottles that are dumped in the UK each year. Julia Platt Leonard meets the innovators who have ditched the plastic and want us to eat water instead.

I've slurped water. I've guzzled it. I've sipped it. But I've never eaten it. That changed when I tried my first Ooho. Ooho – or edible water – is the brainchild of Pierre Paslier and Rodrigo Garcia Gonzalez, who wanted to create an alternative to plastic bottles, the ones many of us buy every day and toss away. Their ingenious solution is an edible, seaweed-based membrane that holds water.

Quirky? Yes, but also very smart. Their company Skipping Rocks Lab just launched a funding initiative through Crowdcube and raised £850,000 in only 72 hours, smashing their initial target of £400,000.

The two met while studying for a masters in innovation design engineering, a programme offered jointly by Imperial College and the Royal College of Art in London.

They set themselves the Herculean task of finding an alternative to plastic, using natural materials. "The problem is that a plastic cap or plastic bottle will take 700 years to decompose, so there is a complete mismatch between how long they are going to be used and how long the environment is going to take to decompose them," Paslier says. "Our goal is to match the actual time it's going to take to consume, with the right packaging."

They found inspiration in an unusual place. "The starting point for the exploration was to look at fake caviar – fish balls that are actually made of alginate which is an extract from brown seaweed," he says. Seaweed, it turned out, ticked a lot of boxes. It's readily available around the world so in the future their edible water can be produced locally rather than being shipped long distances, dramatically reducing the carbon footprint. Seaweed is also fast growing. "It's something that is so renewable that it makes you wonder why we are using precious materials when nature provides materials that are so renewable."

The group Recycle Now estimates that the average UK household uses around 480 plastic bottles a year but only recycles around 270, so nearly half are never recycled. That translates into nearly 16 million bottles dumped every year across the country, winding up in landfills or the ocean. Of course, not all of those are water bottles but they are a big culprit. The problem is compounded, Paslier says, since it takes a lot of natural resources to make plastic bottles in the first place, making them costly in more ways than one.

On paper, the two perhaps weren't the most likely candidates to tackle the huge issue of plastic bottles. "His background is architecture," says Paslier, "and mine is in mechanical engineering, so we weren't really supposed to be the ones to bring this sort of innovation to the world."

Currently the two are focusing specifically on a replacement for the small, single-serve bottles you buy when you're on the go. Paslier says this accounts for about 30 per cent of the bottled water market in the UK. They're developing the technology so edible water can be made on the spot so you could walk into a place such as Pret A Manger, Leon or Starbucks, pick up something to eat and grab an Ooho too. "We're actually talking to some of these chains to provide the technology so they can produce Ooho and sell them as they would fruit. And that's really where we see a lot of potential in the next few years."

Their goal isn't to add another brand of water to an already crowded market but instead to offer a new packaging technology. They envision a range of products so, for example, there would be one version with a very thin membrane suitable to hand out to runners in a race. For the retail market, there would be another with a thicker outer layer that you could peel off and discard, making it more durable and hygienic.

Coming up with the idea for Ooho was the first step but turning it into a reality was another. "It's definitely an uphill battle, I'm not going to lie. But at the same time, it's easier than ever for ideas like this to become a reality," Paslier says, but help from a programme called Climate-KIC, the EU's climate innovation initiative, was invaluable in getting started.

It all makes for a business success story but listening to Paslier, the real joy comes from helping to solve a tricky problem. "What's interesting is that we don't necessarily want to block other people from exploring the potential of seaweed," he says. "We want to be part of this ecosystem so of course we protect our technology, but I think there's room for a lot of people to solve these problems."

And what's it like to eat/drink an Ooho? It's surprisingly easy (I had worried that it would burst like a water balloon) and you can eat the seaweed membrane or discard it. "It's fun, it's different and provides a completely guilt-free experience," says Paslier. Not bad for a drink of water.

14 April 2017

⇨ The above information is reprinted with kind permission from *The Independent*. Please visit www.independent.co.uk for further information.

What happens to our recycling?

Ever wondered what happens after your recycling bins are collected or you drop something off at the recycling centre? We explain how your recycling becomes a valuable resource.

Recyclable items are predominantly collected from your home in two ways. There are kerbside 'sort' schemes where recyclables are sorted into their respective materials on the lorry at the kerbside; and co-mingled collections where all your recyclables are put into one compartment on the lorry before being taken to a Materials Recovery Facility (MRF) and sorted.

At the MRF, all the mixed recycling is sorted and separated into different types of materials by hand or machine (or both) before being sent to manufacturers who make it into new products.

Once collected and sorted, recycled materials become valuable commodities in the worldwide market.

What is recycled in the UK?

There are many recycling plants in the UK, reprocessing million of tonnes of material every year.

⇨ All of the newsprint manufactured here in the UK is now made from 100% recycled paper.

⇨ All of the organic (garden and kitchen) waste we collect is recycled here, usually quite close to where it is collected.

⇨ Over 80% of the glass collected for recycling is used in the UK, the majority of it to make new glass bottles and jars.

There is an ever-increasing range of high-quality products that are made in the UK from recycled materials. To find out what happens to the things you recycle watch our short animations and read the fact files.

Does the UK export any recyclable material?

Countries such as China are prepared to pay high prices for recyclables such as waste plastic; mainly because they do not have readily available sources of virgin materials (no indigenous forests or oil supplies) and they have a large manufacturing industry that requires these products.

Even though exporting our recyclables means a bigger recycling loop because recyclable materials are transported

further, it is still a better environmental option than using virgin, raw materials.

⇨ It minimises the need to use our natural resources such as oil. Using recycled materials significantly reduces energy use and carbon emissions during the manufacturing process.

⇨ The transport impacts are reduced because the materials are transported in container ships returning to China after bringing the goods to the UK.

⇨ It means those materials are not being landfilled.

⇨ The above information is reprinted with kind permission from Recycle Now. Please visit www.recyclenow.com for further information.

© Recyclenow 2017

RECYCLE
SEPARATE WASTE MATERIALS
COMPOST
CHOOSE RECYCLABLE!

REUSE
USE THINGS MORE THAN ONCE
REPAIR
REGIFT!

REDUCE
AVOID WASTE!
BUY LESS
CONSERVE WATER

Five easy steps to reduce waste

Stop unwanted mail

To stop receiving unsolicited advertising mail, contact the Mailing Preference Service and Royal Mail.

Be aware when giving your name and address; ensure you indicate you don't want your details to be used for future promotions or passed to other marketing agencies.

Think before you shop

Before you shop, think about how you could reduce your waste. For example, you could buy things with less packaging, or second-hand items.

Can you buy something that will last longer rather than a disposable item? Rechargeable batteries are cheaper than disposables, and they can be used again and again – and using real crockery and cutlery for your parties instead of disposable ones can make a real difference.

To avoid taking home more carrier bags, choose reusable shopping bags: keep them in the car boot or pop a folded cotton bag in your handbag or pocket so you always have one to hand.

Your rubbish is another person's treasure!

Contact your local Furniture Re-use Network – they distribute unwanted furniture and household goods to those in need. Or advertise your old furniture in your local paper or on websites such as Freecycle or Freegle.

Having a clear out? Donate good quality clothing, books and unwanted gifts to charity shops. Alternatively, use the clothing banks at your local recycling centre.

Avoid wasting food

Every year in the UK we throw away £12.5 billion worth of good food, costing the average family almost £60 a month.

To find out useful tips on the storage of food, tantalising recipes and advice on portioning and meal planning, visit Love Food Hate Waste.

Re-paint

Donate left-over paint to your local charity. To find your nearest scheme visit Community Repaint.

What more can I do?

It's not just our household waste that affects climate change. There are lots of other things you can do to help reduce the amount you waste and your impact on the environment.

The food we throw away, the homes we live in and the cars we drive all emit carbon dioxide emissions – the main greenhouse gas that causes climate change.

The Department of Energy & Climate Change (DECC) recently reported that over 40% of CO_2 emissions in the UK come directly from what we do as individuals; for example, heating and using electricity in our homes and driving vehicles. Flights are the next biggest source of individuals' CO_2 emissions. (For more information, see the DECC (2013) official statistics: *Final UK greenhouse gas emissions*.)

Here's a rundown of other ways you can reduce your carbon footprint and some of the organisations that can help.

Reduce the food you waste

Around a staggering £12.5 billion worth of food is thrown away in Britain annually, if we stopped wasting good food we could avoid around 17 million tonnes of carbon dioxide equivalents being emitted each year, the same as taking one in four cars off of UK roads.

Love Food Hate Waste provides easy tips on reducing the amount of food you waste, useful information on ideal portion sizes, ready-made meal plan recipes and handy shopping lists – plus much more to help you reduce the amount of food you throw away!

Save energy in your home

The Energy Saving Trust can help you to make some simple energy saving home improvements, save money on your energy bills, improve the comfort of your home and help fight climate change.

Waste less water

Did you know that waiting for the tap to run cold when filling a glass with water can waste more than ten litres of water a day?

Waterwise provides handy tips on how to reduce the amount of water you waste and make your home more water efficient.

Do the green thing

Green Thing is an online community for anyone who wants to be a bit greener and needs some advice on how to do this. It focuses on seven things you can do – and enjoy doing – to live a greener lifestyle.

⇨ The above information is reprinted with kind permission from WRAP. Please visit www.recyclenow.com for further information.

© WRAP 2017

New packets could use nano-tech to tell you when your food is about to go off

Prevent that dreaded moment when it's too late.

By Sophie Gallagher

There is nothing more annoying than making the perfect cup of tea, only to take a mouthful and discover your milk has spoiled.

And as people who seem unable to sufficiently organise a fridge to stop this happening on a fairly regular basis, we could really do with some assistance.

Now plans for nanotechnology to be integrated into food packaging could help stop this, and help with other first-world problems, such as our wine being the wrong temperature to drink.

The groundbreaking tech, from researchers at Trinity College Dublin, integrates 2D printing with new nano-materials, such as graphene, phosphorene and molybdenite to create nanosheets.

These printed sheets could then have electrical functionality such as a countdown clock for your food's use-by date that then sends a text message to your phone to remind you to go to the shop.

Professor Jonathan Coleman said: "In the future, printed devices will be incorporated into even the most mundane objects such as labels, posters and packaging."

The team also suggest it could be used for a more secure generation of banknotes, or e-passports.

"Printed electronic circuitry (constructed from the devices we have created) will allow consumer products to gather, process, display and transmit information," said Coleman.

Printable electronics have developed over the last 30 years mainly based on printable carbon-based molecules.

While these molecules can be turned into printable inks, such materials are unstable and have well-known limitations because they cannot necessarily conduct, or insulate well enough for circuits.

Not only will this new technology stop us from ruining a brew, but could be used to address the level of food waste in the UK.

Almost 500 million pints of milk are wasted each year, according to food waste reduction charity Wrap, with 20% of those based on labelling on the packet.

7 April 2017

⇨ The above information is reprinted with kind permission from The Huffington Post UK. Please visit www.huffingtonpost.co.uk for further information.

Waste not, want not: the circular economy turning waste into a resource

Imagine a world with minimal waste, where practically everything is upcycled and re-used. Where there's no more landfill and mountainous food waste is used for biogas production. Imagine a resource-efficient society with zero waste, where waste is no longer a burden, but a profitable resource. Sounds wonderful, doesn't it?

Turning waste into a resource has far-reaching benefits, both for the environment and the economy, and everyone can benefit, from corporate organisations to SMEs (small and medium-sized enterprises), local authorities to agricultural sectors, and yes, individuals too.

The good news is, we're slowly making steps towards using resources in a more sustainable way, driven by a new action plan that incorporates waste management into the strategies of both products and services. It's known as the circular economy and is set to revolutionise the way we tackle waste.

What is the circular economy and why is it beneficial?

First proposed as a concept back in the 1970s, it's an alternative to the conventional and unsustainable linear manufacturing process, where the majority of products are made, used and then disposed of at the end of their life with very little materials recycled or re-used.

The idea is to keep raw materials and products in circulation for as long as possible, making secondary resources the preferred option over primary raw materials. A circular economy achieves this by emphasising the use of the Waste Hierarchy: prevention, repair, re-use, recycling and recovery, rather than disposal.

The benefits include:

⇨ Reduced environmental impact of production and consumption

⇨ Less waste

⇨ A more competitive economy

⇨ A practical solution to our resource problems

⇨ Improved resilience to changing markets.

New job opportunities

The concept only started to gain significant weight from 2010, when former sailor Dame Ellen MacArthur became involved. She established the Ellen MacArthur Foundation to champion an economy that is restorative and regenerative. Six years on, global corporations, world leaders and celebrities have now become vocal supporters of its principles.

Helping Europe's transition towards a circular economy

The UK Government and the European Union are very much in favour of this approach, and have revised legislative proposals and waste reduction targets to help make this transition happen. They include reducing landfill to a maximum of 10% of all waste by 2030 and targets for recycling packaging and municipal waste of 75% and 65%, respectively by 2030.

Their proposals are ambitious and it won't happen overnight. To achieve these targets, we will be required to make significant changes in the way that we design both products and packaging, and how we decide to handle and manage them and all the waste that our products and services produce. We will need to rethink how we implement new business models,

new technologies for recycling, and of course, changes in consumer behaviour.

At the heart of this revolution is the resource management industry. To make the transition will require a new way of thinking that goes beyond recycling by embracing prevention, re-use and refurbishment practices. Products will need to be designed for eventual recycling, once the preferred waste management options presented in the Waste Hierarchy (prevention and re-use) have been considered and used.

A circular economy will help to deliver a much more efficient and competitive UK economy, creating more jobs, generating sustainable growth and an energy and resource efficient community. It hopes to achieve this by revolutionising self-sufficiency and the implementation of industrial symbiosis; where companies realise that what is waste to them could be used as a raw material by someone else, thus preventing the need to enter new raw materials into the supply chain, helping to reduce our ecological and carbon footprint in the process.

The different phases of the circular economy

A circular economy can be broken down into several stages: production, consumption, waste management and secondary raw materials.

Production and consumption

Under the Ecodesign Directive, the European Commission will support the introduction of smart production design and improved processes at production stage, which will help save resources, make waste management more efficient and create new business opportunities. To ensure producers make products that can be easily recycled and reused, financial incentives will be introduced.

The environmental impact and financial savings are key to encouraging the uptake of sustainable products and services. Re-using and repairing products to extend their life has both financial gains and reduced waste gains. Products will need to be durable and repairable if they are to deliver substantial benefits to consumers.

Businesses will also benefit from resource-efficient, recyclable products. Products and services that are supportive of the circular economy will lead to an increase in new business opportunities and new jobs, particularly in the design and repair sectors.

Waste management and secondary raw materials

Who'd have thought that waste could become a valuable resource? In a circular economy, secondary raw materials have huge potential, reducing environmental impact and production costs.

Trading in these materials is an underdeveloped market; it requires improvements to waste management practices and improved quality standards. Currently only 25% of waste material from SMEs is resold, climbing to 45% for larger companies.

Waste management and environmental impacts

Did you know that just 40% of waste produced by EU households is recycled? Around 600 million tonnes of lost waste materials could potentially be re-used or recycled.

New waste management proposals will also help to reduce greenhouse gas emissions. By 2035, more than 500 million tonnes of greenhouse gas emissions could be prevented: indirectly through the recycling of materials which would usually be extracted and processed, and directly by reducing landfills leading to a cut in associated emissions.

23 March 2016

⇨ The above information is reprinted with kind permission from Environment Energies. Please visit www.environmentenergies.co.uk for further information.

Key facts

⇨ By 2025 there will be 1.4 billion more people living in cities worldwide, with each person producing an average of 1.42kg of municipal solid waste (MSW) per day – more than double the current average of 0.64kg per day. (page 1)

⇨ The top producers of municipal solid waste (MSW) are small island nations, including Trinidad & Tobago (14.40 kg/capita/day), Antigua and Barbuda (5.5kg) and St Kitts and Nevis (5.45kg), Sri Lanka (5.10kg), Barbados (4.75kg), St Lucia (4.35kg) and the Solomon Islands (4.30kg). Guyana (5.33kg) and Kuwait (5.72kg) also scored highly. (page 1)

⇨ The UK recycling rate for 'waste from households' was 44.3 per cent in 2015, falling from 44.9 per cent in 2014. This is the first time the rate has fallen since it began in 2010, though the 2015 figure still represents the second highest annual value on record. There is an EU target for the UK to recycle at least 50 per cent of household waste by 2020. (page 2)

⇨ The UK generated 202.8 million tonnes of total waste in 2014. Over half of this (59.4 per cent) was generated by construction, demolition and excavation, with households responsible for a further 13.7 per cent. (page 2)

⇨ According to a 2015 Government consultation, various sources have estimated the cost of fly-tipping on private land as between £50–£150 million a year. (page 5)

⇨ The penalties for fly-tipping set out in the Environmental Protection Act 1990 were increased through the Clean Neighbourhoods and Environment Act 2005. It is now a criminal offence punishable by a fine of up to £50,000 or 12 months imprisonment if convicted in a Magistrates' Court. The offence can attract an unlimited fine and up to five years imprisonment if convicted in a Crown Court. (page 6)

⇨ In 2014, over 40 million tonnes of electronic waste were generated around the world, up to 70% of which was discarded in China. (page 8)

⇨ 70% of the e-waste that is processed at recycling units in New Delhi are exported or dumped by developed countries. Delhi is the leading processing center of e-waste in the country: about 25,000 workers illegally refurbish up to 20,000 tonnes of e-waste annually. India generates about 3.5 million tonnes of e-waste every year, and imports another 50,000 tonnes from countries including the US, Canada, EU, Australia and countries in the Middle East. (page 9)

⇨ In 2014, Glastonbury Festival recycled 54% of the waste produced onsite.

⇨ Reading Festival only managed to recycle about a quarter of their total waste in 2013 and 2014. In fact, there was a 205-tonne increase in waste produced on site between 2013 and 2014, combined with a 2% drop overall in recycling. 2013 saw 433 tonnes of waste and 169 tonnes recycled, and 2014 saw 596 tonnes and 214 tonnes recycled. (page 11)

⇨ In the UK, we throw away seven million tonnes of food and drink from our homes every year, the majority of which could have been eaten. Wasting this food costs the average household in the UK £470 a year. (page 17)

⇨ Tesco estimates it produced 59,400 tonnes of food waste last year across all its UK retail operations, which includes its stores, petrol stations and food depots. It's the only supermarket to publish figures on its food waste. The figure doesn't include waste produced by Tesco's suppliers, or food bought from Tesco and wasted by households. (page 18)

⇨ Statistics from the Waste and Resources Action Programme (WRAP) show household food waste in the UK increased 4.4 per cent between 2012 and 2015, despite a target to cut household waste 5 per cent by 2015. That brings the amount of food sent to landfill in 2015 to 7.3 million tonnes, costing UK families £700 per year. (page 19)

⇨ E-waste is one of the world's fastest growing waste streams. We generate over 40 million tonnes annually. (page 26)

⇨ Trillions of tiny pieces of plastic are accumulating in the world's oceans, lakes and estuaries, harming marine life and entering the food chain. (page 30)

⇨ Around a staggering £12.5 billion worth of food is thrown away in Britain annually, if we stopped wasting good food we could avoid around 17 million tonnes of carbon dioxide equivalents being emitted each year, the same as taking one in four cars off of UK roads. (page 36)

⇨ The Department of Energy & Climate Change (DECC) recently reported that over 40% of CO_2 emissions in the UK come directly from what we do as individuals; for example, heating and using electricity in our homes and driving vehicles. Flights are the next biggest source of individuals' CO_2 emissions. (For more information, see the DECC (2013) official statistics: *Final UK greenhouse gas emissions*.) (page 36)

⇨ Almost 500 million pints of milk are wasted each year, according to food waste reduction charity Wrap, with 20% of those based on labelling on the packet. (page 37)

⇨ Currently only 25% of waste material from SMEs is resold, climbing to 45% for larger companies. (page 39)

Biodegradable waste

Materials that can be completely broken down naturally (e.g. by bacteria) in a reasonable amount of time. This includes organic materials such as food waste, paper waste and manure, which can be composted, as opposed to items such as plastic bottles that would take thousands of years to break down naturally.

Circular economy

An economy in which we keep resources for as long as possible, then reuse/regenerate products at the end of their life cycle.

Fly-tipping

Sometimes referred to as `sneaky dumping` or `dumping on the fly`, fly-tipping is the illegal dumping of waste in inappropriate areas. People usually do this so they don't have to pay for bulky items to be collected and removed. Fly-tipping is unsightly and poses a threat to the environment and human health.

Incineration

A method of disposing of waste by burning it into ashes. Incineration reduces the amount of waste that is sent to a landfill and can even convert waste into energy. However, there are concerns about the environmental impact of incinerators (air pollution, toxic waste, etc.).

Landfill

A type of waste disposal in which solid waste is buried underground, between layers of dirt. Biodegradable products will eventually break down and be absorbed into the soil: however, non-biodegradable products such as plastic carrier bags will not break down (or will do so very, very slowly).

Litter

Rubbish that has been discarded and left lying around rather than disposed of properly. Littering is a crime and people can be fined on-the-spot up to £80.

Microbeads

Solid plastic particles, less than 5mm in size. Often used in cosmetics, there is call for them to be banned as they enter the ocean and can be damaging to marine life.

Recycling

The process of turning waste into a new product. Recycling reduces the consumption of natural resources, saves energy and reduces the amount of waste sent to landfills.

The three Rs/Waste hierarchy

The three Rs of recycling are Reduce, Re-use and Recycle. This refers to reducing the amount of waste you make, re-using materials rather them throwing them away (for example, glass milk bottles delivered to your doorstep get used again) and recycling materials by breaking them down and remaking them into something else (plastic drink bottles could be melted down and be made into a plastic chair).

Throwaway/throw-away society

A society where rather than re-using or recycling something, people just throw it away. This is strongly influenced by consumerism, the increased consumption of goods.

Waste

Anything that is no longer of use and thrown away. Each year the UK generates approximately 290 million tonnes of waste, which has a damaging effect on the environment.

Waste Electrical and Electronic Equipment (WEEE) directive

This EU Directive aims to reduce the amount of waste from electrical and electronic equipment and increase its re-use, recovery and recycling. Electrical items cannot just be throw away, they have to be disposed of in a particular manner. This includes large and small household appliances (e.g. washing machine or kettles), lighting equipment, electric tools, toys, etc. Before the implementation of the WEEE directive, waste like this was simply thrown into the bin, but now producers and supplies of electrical goods need to offer their customers a way to dispose of their waste properly (known as `takeback`).

Zero waste

A plan to promote the idea of recycling and re-using materials rather than just disposing of them. The aim is to reduce the amount of waste sent to landfills.

Assignments

Brainstorming

In small groups, discuss what you know about waste and recycling.

- Which countries produce the most waste?

- How is our waste disposed of (in the UK)?

- What kind of waste goes to landfill?

Research

⇨ Read the article on page one, *Which countries produce the most waste?*, and choose a country to research further. Find out how much waste your chosen country produces, what kind and how they dispose of it. Write your findings in bullet-point form, using no more than one side of A4.

⇨ Read the articles in chapter two. Many of them introduce new and exciting ideas for tackling waste. In pairs, do some further research online and choose a revolutionary solution to share with the rest of your class.

⇨ Find out which European cities are best at dealing with their waste, feed back to your class.

⇨ List different types of waste management (e.g. incineration, landfills, recycling, zero waste, etc.) and create a list of pros and cons. Which method would you recommend?

Design

⇨ Choose one of the articles from this book and create an illustration that highlights the key themes of the piece.

⇨ Design a poster that illustrates the waste statistics from the article on page 2.

⇨ Design an infographic that demonstrates the concept of a circular economy.

⇨ Design a storyboard for a Youtube video which will encourage people to reduce their household food waste. You could plan to use actors or animations. Include some information about why it is important for food waste to be reduced and some tips on how to do so.

⇨ Design a leaflet that will be distributed by your local council to draw attention to the negative effects of fly-tipping.

⇨ Create a campaign that will raise awareness of electronic waste amongst people your age. What kind of campaign would be most successful? Television, radio, web or posters? Produce a campaign plan and include sample designs, scripts or storyboards.

Oral

⇨ Imagine that you work for a company who's core product offering is re-usable coffee cups. In pairs, create a PowerPoint presentation in which you aim to persuade a chain of coffee shops to stock your re-usable cups. You should focus heavily on the environmental impact of disposable coffee cups.

⇨ In small groups, discuss the pros and cons of incineration and landfills.

⇨ As a class, stage a debate in which one half of you argues that fly-tipping is completely excusable because some people simply can't afford the nominal fees charged to use local tips. The other half of the class should argue that it is never, under any circumstances, acceptable. Try to think about the argument from the point of view of a farmer who has to regularly dispose of waste that is dumped illegally on his land and a low-income family who need to dispose of a large electrical item but can't afford to pay the £5 charged by their local tip.

⇨ Do we have a moral duty to recycle our rubbish? Discuss this question with a partner and make notes. Feed back to the rest of the class.

Reading/writing

⇨ Write a letter to the organisers of a music festival in your local area, explaining why it is important they try to minimise waste at the festival and suggesting ways in which they might do this.

⇨ Write a report on the use of microplastics in cosmetics for your school or local newspaper. Use the articles in this book and online research, and write at least 500 words.

⇨ Read the article *Edible water…* on page 34 and write a blog post exploring whether you think this is a good idea or not.

⇨ Read the article *Five easy steps to reduce waste* on page 36 and add at least four more things to the list.

⇨ Research a charity or organisation that tackles the issue of food waste in the UK. Write a report explaining what your chosen organisation does and encouraging people to get involved. For example, you could look at Food Cycle, The Gleaning Network or The Sustainable Restaurant Association.

⇨ Write a list of items that you or your family have thrown away in the last couple of months. Try to think of some alternative uses for these things. For example, that old t-shirt you threw away could have been re-used and turned into a pillowcase.

Acknowledgements

The publisher is grateful for permission to reproduce the material in this book. While every care has been taken to trace and acknowledge copyright, the publisher tenders its apology for any accidental infringement or where copyright has proved untraceable. The publisher would be pleased to come to a suitable arrangement in any such case with the rightful owner.

Images

All images courtesy of iStock except p.6 © Alan Stanton, p.17 © Aranxa Esteve, p.15 © Freestocks, p.16 © Griffin Keller, p.20 © Brandon Morgan, p.21 © Tamara Bellis, p.24 © Amy Morgensen, p.30 © Alysa Bajenarn, p.33 © MorgueFile, p.36 © Thomas Gaustae and p.38 © Lacey Williams.

Illustrations

Don Hatcher: pages 13 & 29. Simon Kneebone: pages 10 & 39. Angelo Madrid: pages 4 & 28.

Additional acknowledgements

Editorial on behalf of Independence Educational Publishers by Cara Acred.

With thanks to the Independence team: Mary Chapman, Sandra Dennis, Jackie Staines and Jan Sunderland.

Cara Acred, Cambridge, May 2017